QUANTUM BRAIN DYNAMICS AND CONSCIOUSNESS

ADVANCES IN CONSCIOUSNESS RESEARCH

ADVANCES IN CONSCIOUSNESS RESEARCH provides a forum for scholars from different scientific disciplines and fields of knowledge who study consciousness in its multifaceted aspects. Thus the Series will include (but not be limited to) the various areas of cognitive science, including cognitive psychology, linguistics, brain science and philosophy. The orientation of the Series is toward developing new interdisciplinary and integrative approaches for the investigation, description and theory of consciousness, as well as the practical consequences of this research for the individual and society.

Volume 3

Mari Jibu and Kunio Yasue

Quantum Brain Dynamics and Consciousness

QUANTUM BRAIN DYNAMICS AND CONSCIOUSNESS

AN INTRODUCTION

MARI JIBU

KUNIO YASUE

Notre Dame Seishin University, Okayama

JOHN BENJAMINS PUBLISHING COMPANY
AMSTERDAM/PHILADELPHIA

The paper used in this publication meets the minimum requirements of the American National Standard for Information Sciences – Permanence of Paper for Printed Library Materials, ANSI Z39.48-1984.

DOI 10.1075/aicr.3

Cataloging-in-Publication Data available from Library of Congress:
LCCN 95035756

ISBN 978 90 272 5123 7 (EUR) / 978 1 55619 183 1 (US) (PB)
ISBN 978 90 272 8492 1 (E-BOOK)

John Benjamins Publishing Company · www.benjamins.com

Dedicated to the memory of
Dr. Hiroomi Umezawa

CONTENTS

Prologue

Since the authors of this book first became aware of the vital scientific adventure of those who created quantum theory at the beginning of this century, we have felt chagrin that we were not born at that time. How fantastic and exciting it would have been to participate in the development of this revolutionary theory of energy and matter. Niels Bohr, Albert Einstein, Louis de Broglie, Erwin Schrödinger, Werner Heisenberg, and Wolfgang Pauli penetrated the fundamental order of nature and gleaned deep insights that changed both scientific and philosophical world views. Almost a century has passed, but quantum theory still provides the fundamental framework for investigating the laws of nature from a physical point of view. Those of us born after the quantum revolution would feel privileged to have participated in that truly exciting scientific adventure.

Almost all physicists have reconciled themselves to the task of refining and enlarging their knowledge of the material world. Only a few have either resisted the fundamental principles of quantum theory and remained focused on the issue of reality or sought another fundamental issue to research. Although many research fields in science might be regarded as basic, at least by the researchers concerned, strictly speaking there are only four basic areas of research (Werbos 1993):

- Reality: What is matter?
- Universe: What is the cosmos?
- Mind: What is mind?
- Life: What is life?

Quantum theory answered the first question. The second issue is in the throes of an exciting revolution in which Einstein's general theory of relativity is being reformulated within the realm of quantum theory, thus giving rise to a new picture of the universe in quantum cosmology. The third and fourth areas of research are, indeed, the last terra incognita of science.

One of the authors (Kunio Yasue), remaining in the first basic area of research, has sought a deeper, more fundamental law of nature underlying that of quantum mechanics and quantum field theory. He has developed his own way of understanding the order of nature. Initially, two fundamental points of view pervaded his thinking, those proposed by Hideki Yukawa and Edward Nelson, respectively, in the 1960s. Within the realm of Yukawa's theory of elementary domains, Yasue perceived quanta as nothing but energies localized in and traveling between fundamental elements of space and time called elementary domains; there the quanta manifest random walks, that is, movements governed by chance or accident. A random walk can be investigated by means of probability theory in mathematics. Then, at the point where such fundamental elements of space and time become infinitely small, quanta draw continuous but random trajectories and may be well described by Nelson's probability theoretical formulation of quantum mechanics. At that level, Yasue has found the order of nature in terms of Hamilton's principle of least action, which previously had seemed untenable in quantum mechanics.

The other author (Mari Jibu) has been struggling with the third basic research issue: What is mind? Much inspired by the transcendental philosophy of Ralph Waldo Emerson and the doctrine of the Upanishads (the sacred writings of Hinduism), she looked for a sound scientific basis for understanding what mind is. First she encountered Karl Pribram's theory of neural holography in which memory and consciousness were assumed to be carried by unknown nonlocal processes similar to those appearing in lasers and holography. In the 1960s, many psychologists and philosophers supported the idea of neural holography proposed by Pribram; however, since then it has been greatly underestimated by most neuroscientists. The criticism from those unfamiliar with quantum concepts was that neither lasers nor holography could exist in brain tissue. They laughed and said, "There is no room for lasers in the brain to realize a holography, Karl." We should not blame those neuroscientists of the 1970s and 1980s, for they were not knowledgeable of supersonic holography and electron wave holography, which are now widely known.

Despite the tendency of the neuroscientific community to disregard Pribram's theory of neural holography, Mari Jibu had a keen insight into the intrinsic value of neural holography twenty years after Pribram's first proposal. In 1987, after graduating from Notre Dame Seishin University with a degree in English and American literature, Mari Jibu met with Kunio Yasue and asked him to teach her the essence of quantum theory so that she might understand Pribram's thought. Her medical studies in anesthesiology have also contributed to the development of this theory.

This book is the story of the collaboration between the two authors. It has evolved in a two-fold manner, beginning with a person-to-person instruction in the basics of quantum mechanics and quantum field theory and fanning out into an investigation of brain functioning, consciousness, and the emergence of mind—self-reflective consciousness—as revealed through quantum theory. The final result of their work is the quantum theory of the brain presented in this volume. Now neither of the authors thinks of himself or herself as having missed the experience of a true scientific adventure. Rather, they shed tears of gratitude for their chance to be in the tide of a quantum revolution in the two remaining areas of research: mind and life.

The entire volume is divided into three parts. The first part is devoted to an exposition of the person-to-person instruction in the basics of quantum mechanics and quantum field theory. Although it might belong personally to the authors, they strongly believe that it might be valuable to open the instruction to those interested in physics and quantum theory who are willing to engage in this challenging scientific inquiry. The second part is devoted to an exposition of the conventional theories of brain functioning, including the classical theory, the physiological theory, and the dynamical theory. The third part is a multi-faceted investigation into the issue "What is mind?". This question, from the point of view of quantum theory, is necessarily connected to another fundamental question: "What is life?" Two appendices are included for the reader's convenience: Appendix A is a brief introduction to a new world view derived from modern quantum field theory by Hiroomi Umezawa (1993), which has influenced the present authors' quantum theory of the brain and consciousness. Appendix B is a glossary in which the reader can find the meanings of technical terms distinguished by bold type in the main text. Other technical terms, many of which are in italics, can be reviewed by using the index.

Mind and life are issues of universal and abiding interest. This might be only the beginning of the last scientific adventure in those two remaining areas of basic research. Those interested in the present exposition are cordially invited to join the adventure.

In 1995 it is time to scoff at the old cliché:

> "What is mind?"
> "No matter."
> "What is matter?"
> "Never mind."

June, 1995
Mari Jibu and Kunio Yasue

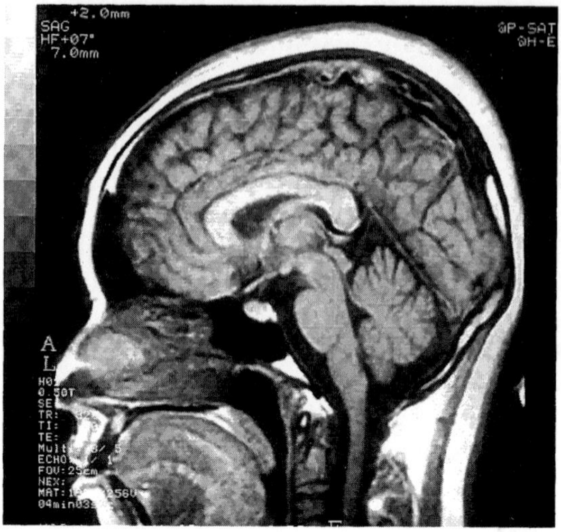

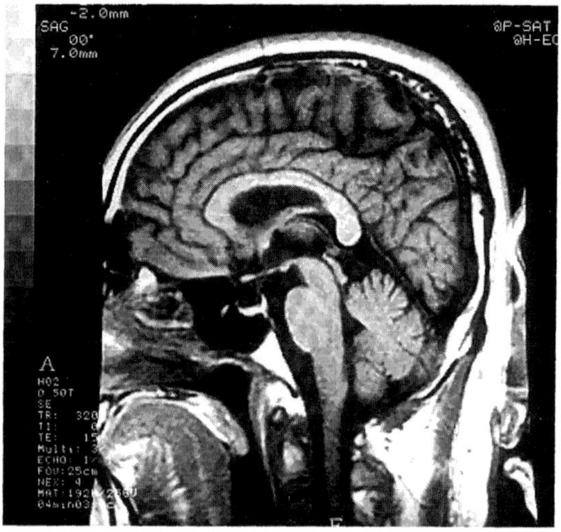

The authors' recent pictures: Mari Jibu (top) and Kunio Yasue (bottom).

Acknowledgments

To adequately express our gratitude to all who have supported our endeavor is surely the most difficult part of this long-term writing project. Many people have been equally helpful to the authors in various aspects of this work. To choose a starting point among so many equally qualified candidates would be impossible. We are fortunate that our language guides the arrangement of the names in a certain order. The authors are, of course, deeply grateful for the blessing of our families—Saeko, Ayako, and Kanako Yasue and Tetsuya, Kiyomi, and Masaki Jibu—whose endless and seemingly thankless tasks have supported our work. Neither of the authors has brothers or sisters, but only parents, spouse, and children. We would like to have had a tender, strong, and intellectual brother. It is Gordon Globus who has kindly played the role of brother to us, stronger, more tender, and more intellectual than our own might have been. The success of the present volume springs not only from the authors' efforts, but also from that of our "brother" living in America, Gordon.

We wish to express sincere appreciation to both Gordon Globus and Maxim Stamenov for their invitation to write the present volume, for the many arrangements they have made, for their valuable suggestions, and for their continual encouragement; to Hiroomi Umezawa and Yasushi Takahashi for their helpful advice and warmhearted guidance on the subject; to Luigi Ricciardi for his assistance and hospitality; to Stuart Hameroff for sharing his idea about consciousness and microtubules; to Sarah and Edward Nelson for their deep interest in our subject; to Paul Werbos and Cole Benton for stimulating conversations; to Giuseppe Vitiello for interesting suggestions and advice on the subject as well as for his recognition of Kunio Yasue as a forerunner of Nelson's probability theoretical formulation of quantum mechanics; to John Watterson for sharing his enthusiasm on water research; and to Donna Skelton for her excellent editing of our writing, including suggestions that helped us to clarify the text for a diverse audience, as well as for her continual encouragement.

Last but certainly not least, we are deeply indebted to Karl Pribram and Katherine Neville, to Scott Hagan and Laura Robinson, and to Jean-Claude Zambrini for making our lives on this planet extremely enjoyable, both scientifically and personally.

We are sad to hear of the deaths of Hiroomi Umezawa and Iain Stuart. We appreciate their significant contributions to science, and we deeply mourn them.

Part One

A Tutorial in Quantum Physics

Chapter 1

Quanta and Fields

Section 1
Mind and Matter

Over the course of history, human beings have developed the view of nature that prevails today, yet we feel ill at ease with it. One cause of the uneasiness is that we have no clear idea what **mind** is. Even **René Descartes**, often called the father of modern science, could not dissolve this uneasiness. He had to put mind outside his impregnable mechanistic view of nature in which the **law of causality** came to the fore. Mind was placed exterior to the **matter** that constitutes nature and was regarded as being in opposition to matter. Mind, thus banished from the realm of modern science, has been taken into serious account almost exclusively in philosophy, epistemology, and the mystical aspect of theology.

The notion that mind is far beyond material existence acquires new meaning in light of quantum physics. Theoretical physicists, who study the nature of matter from the point of view of modern science, often become interested in mind and **consciousness** after they have fully explored matter. Many of the scientists engaged in this inquiry have contributed to the development of a new theory of physics called **quantum theory**, which has changed our view of nature drastically in the twentieth century.

Niels Bohr of Denmark, called the father of quantum theory, was one of those scientists. He suggested that such small amounts of **energy** as were the concern of quantum theory also determined essential features of the thinking process of a human being (Bohr 1958). Then the physicist **David Bohm**, who originated an interpretation of quantum theory called the **ontological interpretation of quantum mechanics** and sought to understand the problem of **nonlocality** inherent in quantum theory (Hiley and Peat 1987), discussed the possibility that changes in states of consciousness are analogous to those of **quantum states** (Bohm 1951).

Preceding Niels Bohr (1958) and David Bohm (1951), an Austrian physicist, **Erwin Schrödinger**, had published a series of papers in 1926 in which he formulated a quantum theory of the mechanics of the fundamental constituents of matter (Schrödinger 1926). His theory has been called **quantum mechanics**. To escape Nazism, he moved to Dublin, Ireland, where he channeled his energy into lectures on essential aspects of life, including mind and consciousness. His famous lecture entitled "WHAT IS LIFE?" led many young physicists to the study of life phenomena, and modern **molecular biology** was born (Schrödinger 1944). Late in life, while residing in a small mountain village in the Kitzbühel Alps, he crossed the valley between mind and matter and, having bridged the two, ended his days in 1961 as one of the world's great physicists.

Schrödinger was in Zürich, Switzerland, when a **wave** concept of quantum mechanics first came to him. At that time, the gifted physicist **Wolfgang Pauli**, who was also in Zürich, contributed much to the development of quantum theory. Pauli discovered the **exclusion principle** of quantum theory, which led to an understanding of the **quantum mechanics of many-body systems** (Pauli 1925). Collaborating with the famous psychiatrist **Carl Gustav Jung**, Pauli also tried to provide a quantum theoretical foundation for two basic concepts of Jungian psychology: the **collective unconscious** and **synchronicity** (Jung and Pauli 1955). He produced several papers formulating various aspects of mind from the point of view of a quantum theory of the **electromagnetic field**, but they have not yet been made public.

Quantum theoretical many-body systems are noteworthy here, for they are known to manifest peculiar characteristics. The most well-known of those phenomena is **superconductivity**, in which electrical resistance vanishes and electric current flows perpetually. The **Josephson effect** is a strange phenomenon in which a **superconducting current** vanishes if a DC voltage is applied to a **superconducting medium** and comes back again if an AC voltage is applied to it. As its name indicates, that phenomenon was predicted theoretically by a young physicist **Brian Josephson**. At the Cavendish Laboratories in England, Josephson has long been investigating the mechanism of consciousness and thought from the physical point of view of quantum theory.

Why, after glimpsing the ultimate aspects of matter through the development of quantum theory, have these people become interested in mind and consciousness and begun new research to acquire a deeper understanding of life? Maybe only we ordinary people think of this as new research. For those who have fully explored the material basis of the natural world, even consciousness and thought constitute the subject matter of physics along the lines of a materialistic world view; that is, consciousness and thought are not separate from matter, though people ordinarily see mind in opposition to matter.

We are going to step into the sub-microscopic world of the human brain to pursue fundamental processes there that give rise to consciousness, **memory** and mind. It is an abyss in our universe, far removed from the world of ordinary reality. We cannot take a step toward it without knowing the laws of quantum theory. We require a minimum knowledge of quantum theory to engage in this pursuit.

Several recently published books and articles have highlighted the relationship between consciousness and **quantum physics** (Bohm 1986; Eccles 1986; Hameroff 1987; Penrose 1989; 1994). The notion of **quantum coherence** is the common key to understanding that relationship, and the authors have attempted to explain quantum theoretical concepts in plain language. Quantum theory, however, is so fragile that no one seems to succeed completely in explaining its essential aspects. The reader needs not only to be familiar with the terminology of quantum theory in plain language but to get the correct knowledge of quantum theory in its original form. It is, indeed, one of the aims of the present volume to provide the readers with a minimum but correct knowledge of quantum theory sufficient to understand its fundamental role in brain functioning.

Section 2
Wave and Particle

We may say that the mathematical and mechanistic view of nature initiated by Descartes in the seventeenth century was later brought to fruition by **Sir Isaac Newton**. His masterful, systematic theory, which explains the motion of every material body from billiard balls to planets, is called **classical mechanics**, and today it still boasts of myriad applications, from small-scale phenomena to galactic events. Collisions of billiard balls, the orbits of space shuttles, and many other daily events are explained by **Newton's law of motion** in classical mechanics.

Once we step into the sub-microscopic world of **atoms** and **molecules**, however, things do not proceed so routinely. Our daily friend, classical mechanics, cannot be applied there. A completely new body of knowledge is required to understand events at the sub-microphysical level. Quantum theory is, indeed, this new system. It not only takes the place of classical mechanics; it also predicts the peculiar behavior of matter that is unlike anything we experience in our ordinary reality.

The investigation of the nature of light is one good example that shows us how difficult it is to understand the dynamics underlying the behavior of matter, as revealed by quantum theory. Newton, who gave birth to classical mechanics, promoted the idea that light should be thought of as a **particle**, like a discharged bullet, although this description failed to accurately account for the **interference** of light. Therefore, **Thomas Young** conducted experiments in 1804 to verify the interference effect of light. He advanced a new theory of light in which it was explained as nothing but a wave propagating in an extremely rarefied medium of "elastic matter" that filled up our universe. Young called this elastic matter "**ether**." Thereafter, this concept of light as a wave propagated for about two centuries and reached its peak of acceptance in scientific circles in the latter half of the nineteenth century.

The **theory of electromagnetism**, formulated by **James Maxwell**, described the nature of light in terms of an **electromagnetic wave**, as distinct from Newton's particle picture of light in classical mechanics. According to Maxwell, the electromagnetic field extending over the whole of space propagates as a wave, and a specific range of **wavelengths** correspond to visible light. Maxwell's theory of electromagnetism was supported by Einstein's **theory of special relativity** early in the twentieth century, and the existence of **fields** in general was proposed on the model of an electromagnetic field. Thus, by taking fields into account as extended mechanical objects, Newton's mechanistic view of nature, which had been generally accepted in the nineteenth century, still seemed sufficient to provide a unified description of the phenomena of matter and light. This is called **classical physics.**

It is interesting that a phenomenon that could not be explained by the wave theory of light had already been discovered at the end of the nineteenth century, even before the completion of classical physics. The discovery pertained to a peculiar phenomenon called the photoelectric effect, in which a metal plate with a negative electric charge loses its charge when irradiated by **ultraviolet light**, while irradiation by **visible light** has no such effect. Note that ultraviolet light has a shorter wavelength than that of visible light. In the metal plate, the negative charge is carried by one of the fundamental constituents of matter called an **electron**, and the loss of electric charge is due to the release of electrons. The intensity of light is given by the energy of a corresponding electromagnetic wave, according to Maxwell's theory of electromagnetism; that is, the energy is proportional to the square of the maximum strength of the fluctuations of the undulating electromagnetic field, or, put another way, the energy is proportional to the square of the **amplitude** of the electromagnetic wave. According to Maxwell's theory, even visible light should be able to force the electrons out of the metal plate, as long as the light is sufficiently intense, because its energy is proportional to its intensity as given by the square of the amplitude. That, however, is not the case.

Albert Einstein, who gained instant fame for his theory of relativity, restored the particle doctrine of light and advanced a new theory of the **light quantum** to explain the photoelectric effect. For this work, he became a Nobel laureate. In his theory, light has a particle-like existence as a **quantum**, that is, an elementary unit of energy. Thus Einstein's theory is called the **quantum theory of light**. Today we call a light quantum a **photon**.

Let us consider the following analogy for the purpose of understanding the photoelectric effect. A game played with glass beads (marbles)—namely, a French "petanque" game for children—will serve our purpose. The children arrange their glass beads in a small circle on the ground. A player tosses a glass bead from a distance, aiming at the beads of other players. Hitting an opponent's marble forces it out of the circle. In this analogy, it is significant that the opponents' marbles cannot be forced out easily if a player tosses his own at a slow speed; however, a marble is easily forced out of the circle if struck by one tossed at a high speed. This empirical law can be well understood in terms of **momentum** or energy within the framework of classical physics. Classical mechanics is properly applied to an ordinary-scale material body, such as a glass marble. A glass marble tossed at a high speed has larger values of momentum and energy than one tossed at a slow speed, so it has a stronger impact upon collision.

In this analogy, the opponents' glass beads (marbles) in the circle are electrons in a metal plate, and the marble tossed by a player is a photon (i.e., a quantum of light). Such a simple picture leads us back to Newton's doctrine of light in the seventeenth century, but it becomes inconsistent with the **principle of the constancy of the velocity of light** in the theory of special relativity. Here the analogy breaks down because a marble can move at various speeds, whereas a photon can only travel at one speed.

Indeed, Einstein developed the **special theory of relativity** starting with the principle of the constancy of the velocity of light, while he also unified the particle and wave properties of light in his idea of a light quantum and, thus, incorporated the paradoxical character of light into physics. Einstein perceived a light quantum to be a cryptic physical form of the electromagnetic field, so it may be considered a particle on one hand and a wave on the other hand.

Classical mechanics does not account for the relationship between **velocity**, momentum and energy in Einstein's quantum theory of light, in which the momentum and energy of a photon (i.e., a light quantum) are proportional to the **wave number** and **frequency**, respectively, of the corresponding wave. The velocity of a photon remains constant due to the principle of the constancy of the velocity of light. Photons move with the speed of light regardless of their various momenta and energies. Furthermore, the intensity of light is thought to be proportional to the total number of photons in the light. In this way, a photon corresponding to ultraviolet light with larger values of wave number and frequency has larger values of momentum and energy than a photon corresponding to visible light with smaller values of wave number and frequency; therefore, a photon corresponding to ultraviolet light can force an electron out of the metal plate. On the other hand, the fact that no electrons are forced out even if high intensity visible light is used to irradiate the plate may be understood analogously to the fact that no marbles can be knocked out even if a player tosses many marbles at slow speed, that is, marbles with smaller values of momentum.

Section 3
Double-Slit Experiment

Einstein's doctrine of the light quantum cast a ray of hope into the chaotic realm of physics that had emerged by the beginning of the twentieth century due to unresolved issues such as the photoelectric effect and the failure of the wave description of light to account for that phenomenon. Quantum theory dealt with those issues. Let us look at the double-slit experiment for an electron, which may be one of the simplest experiments to verify the motion of an electron on the submicroscopic scale. Let's see how quantum theory developed from this famous study. Until 1989, it had not actually been performed and had been regarded as a mere thought experiment due to the technical difficulty of fabricating the real experimental device; however, Akira Tonomura and his colleagues at Hitachi Advanced Research Laboratory succeeded in realizing it recently (Tonomura et al. 1989).

Suppose that we have an electron source on the far left side of a room and a screen to detect electrons on the far right side. Situated between the electron source and the detecting screen is a blackout plate with two parallel slits vertically adjacent to each other. (See Figure 1.) The electrons emitted from the source are arranged to have one and the same value of momentum and energy and to go through the double-slit and reach the screen. By emitting electrons successively from the source with a sufficient time interval between them to allow only one electron at a time to be observed, the position of arrival of each electron is recorded on the screen. There, the distribution of electrons on the screen shows a striped interference pattern similar to the pattern displayed in Young's experiment of the interference of light in the year 1804. (See Figure 2.) The interference pattern reveals the effect that is obtained when two systems of waves reinforce or neutralize each other. That is what had convinced Young of the wave-like character of light. Similarly, the wave-like character of the electron was revealed by the double-slit experiment.

The discovery of the photoelectric effect revealed that light, previously thought of as a wave, actually has a particle-like feature. Einstein's quantum theory of light claimed that light consists of a quantum, a cryptic physical form of the electromagnetic field that can be considered a particle on one hand and a wave on the other hand. The double-slit experiment revealed that the electron, previously thought of as a particle, also has a wave-like property. Quantum theory resulted from such experiments, which ingeniously exhibited the wave-particle duality of the fundamental constituents of matter, specifically the electron and the photon.

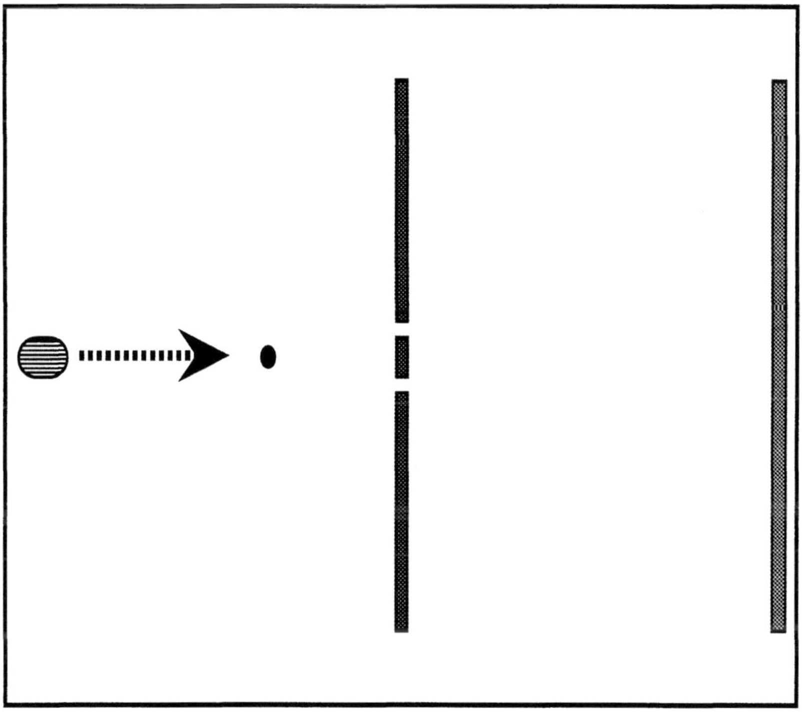

Figure 1. A schematic diagram showing the setup of the double-slit experiment. We have an electron source on the far left side of a room and a screen to detect electrons on the far right side. Situated between the electron source and the detecting screen is a blackout plate with two parallel slits vertically adjacent to each other. The electrons emitted from the source are arranged to have one and the same value of momentum and energy and to go through the double-slit and reach the screen.

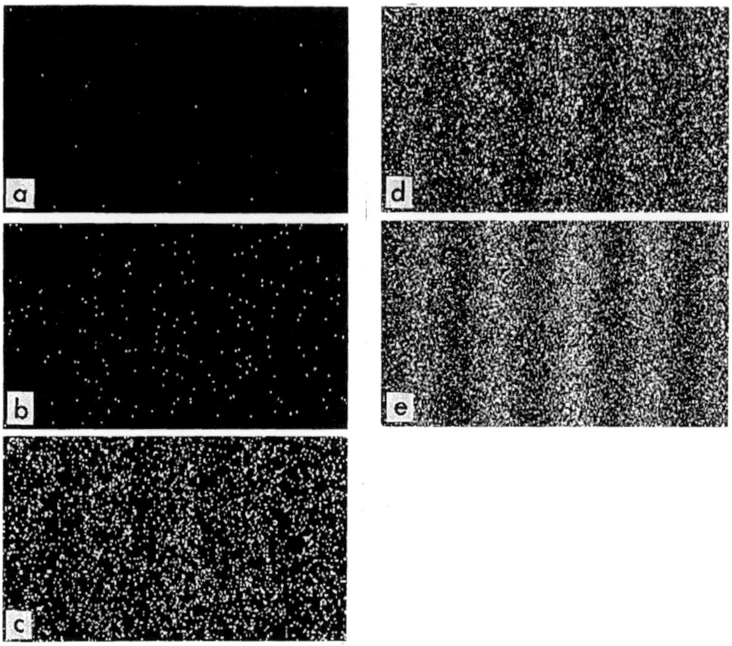

Figure 2. Distribution patterns of electrons recorded on a detecting screen in the double-slit experiment for the electron. While the number of detected electrons remains small (a and b), the interference pattern is vague. However, the interference pattern becomes clearer if the number of detected electrons increases (c, d and e). Photographs taken by Akira Tonomura of Hitachi Advanced Research Laboratory.

The double-slit experiment revealed another interesting fact: the striped interference pattern on the screen vanished if each electron emitted from the source was observed during its passage through the slit. That is, we have to observe an electron just after it has passed through one of the slits to determine which one it has actually passed through. Applying light to an electron coming out of the slits enables us to detect the electron by the reflection of the light. To do this, we have to use light with a wavelength somewhat shorter than the separation between the two parallel slits so that we can clearly judge which one of the slits the electron has passed through. The reason is that the resolution limit of an optical device, such as a microscope or a telescope, to measure the reflected light is approximately equal to the wavelength. Two points with a spatial separation shorter than the wavelength cannot be identified as two different points; they are seen as one and the same point when observed as the reflected light. Recall now the photoelectric effect we have already discussed. According to Einstein's quantum theory of light, light is not only a wave-like entity but also a particle-like entity called a light quantum or photon, and its momentum and energy as a "particle" are proportional to the wave number and frequency of the wave when it is seen as an electromagnetic wave. Therefore, light with a shorter wavelength, that is, one with a larger wave number and a higher frequency, becomes a photon with larger values of momentum and energy. As a result, photons with greater momenta and energies collide with each electron just after it passes through the slit; thus, the motion of the electron is disturbed considerably, and the resulting distribution of electrons on the screen no longer manifests a striped interference pattern. When the photons disturb the motion of the electron, the values of its energy and momentum suffer an uncontrollable change, and so do the frequency and wave number of a "wave" of the electron, due to the wave-particle duality of fundamental constituents of matter. This uncontrollable change of frequency and wave number of the electron "wave" causes broadening and obscuring of the striped interference pattern.

No doubt, the light irradiating each electron for the purpose of observation causes the break in the quantum theoretical interference pattern. Light is a quantum theoretical entity that manifests a particle-like feature when it interacts with matter. Therefore, when light is cast on an electron to observe it, the photons collide with the electron and violently perturb it. This disturbance in the motion of the electron causes the uncontrollable change in the wave number and the frequency of its "wave," so the interference pattern on the screen vanishes. Restated, the collision between the photons and the electrons causes the momentum and energy of each electron to randomly fluctuate from its original value, so the interference pattern of successive electrons reaching the screen disappears. Conversely, if the momentum of a photon is sufficiently small, the disturbance of each electron's wave number and frequency is also reduced, so the striped interference pattern reappears as electrons reach the screen. However, because a photon with smaller momentum corresponds to light with a longer wavelength, we cannot use it to verify which one of the two slits it passes through. (Recall that the resolution limit of light, as observed through a microscope, is approximately equal to the wavelength and that two points with a spatial separation shorter than the wavelength are seen as one and the same point.)

This is the outline of the double-slit experiment.

Section 4
The Double-Slit Experiment for the Photon

The electron has been shown to manifest a wave-like feature in the double-slit experiment. The distribution of electrons on the screen displays a striped interference pattern similar to that revealed in Young's interference experiment for light. We have concluded from this fact that the electron behaves basically like a light quantum, which manifests both wave-like and particle-like properties. In other words, an electron has a quantum nature, as does light.

This question may arise: Is the interference experiment for light (performed by Young) the same as the double-slit experiment for the electron? In the latter, electrons were emitted successively from the source with a sufficient time interval to allow each electron to be tracked individually. In the former, photons were emitted simultaneously from the source. Indeed, the light emitted from the source had sufficient intensity to contain a considerable number of photons. Therefore, it seemed meaningful to perform an updated double-slit experiment for a photon, that is, a light quantum. Such an experiment is as difficult to actually perform as the one for electrons, due to the lack of a photon detector sensitive enough to detect a single photon.

Prior to the success of the double-slit experiment for the electron, Yutaka Tsuchiya and his colleagues at Hamamatsu Photonics Central Research Laboratory had succeeded in performing such an experiment (Tsuchiya et al. 1982). There, photons were emitted successively from a photon source with a time interval sufficient to allow only one photon to be under experiment at any instant. Each photon passed through the double-slit and registered on a screen made of high-sensitivity photon detectors. Indeed, the distribution of photons on the screen showed a striped interference pattern just like the one observed in Young's interference experiment and in the double-slit experiment for the electron. (See Figure 3.)

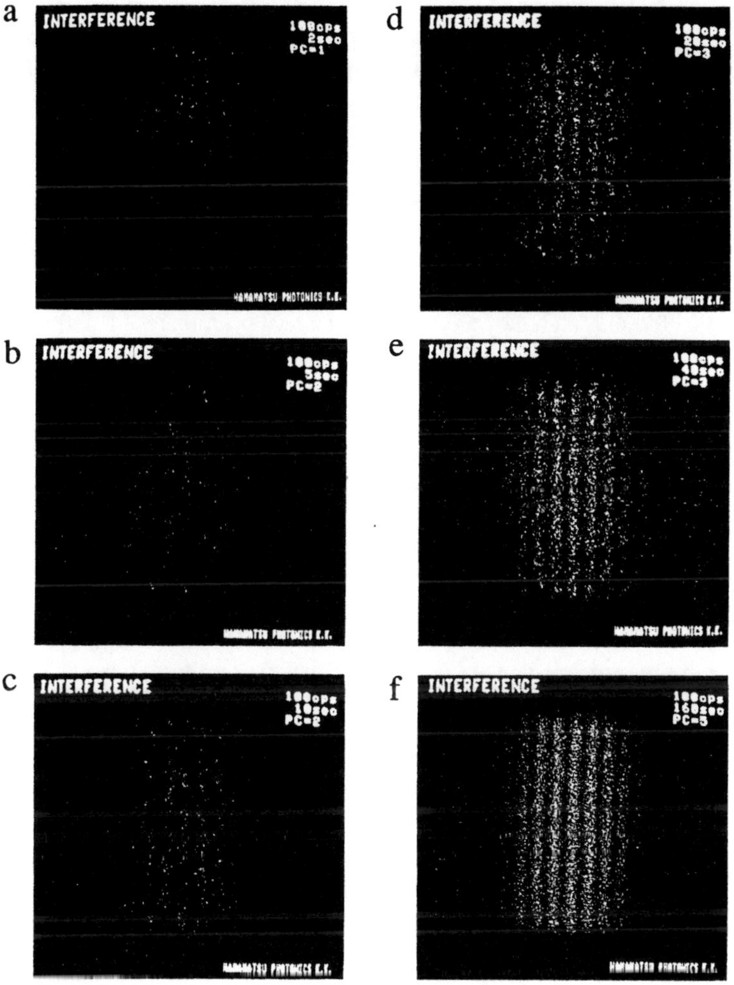

Figure 3. Distribution patterns of photons recorded on a detecting screen in the double-slit experiment for the photon. While the number of detected points of photons remains small (a and b), the interference pattern is vague. However, the interference pattern becomes clearer if the number of detected photons increases (c, d, e and f). Photographs taken by Yutaka Tsuchiya of Hamamatsu Photonics Central Research Laboratory.

The interference pattern does not change essentially even if we perform the double-slit experiment with light including a number of photons simultaneously, as Young did about two hundred years ago. This is not the case for the electron. If a number of electrons are emitted simultaneously from the source, the distribution of electrons on the screen will not typically manifest the striped interference pattern obtained in the case of successive emissions at timed intervals.

Electrons and photons are both quanta, the exact nature of which cannot be understood within the realm of classical physics. They share this fundamental feature as quanta: Both electrons and photons exhibit particle-like and wave-like behaviors. Even though they are both quanta, however, there is a definite difference between them: One is a "boson" and the other is a "fermion." The following will make that distinction clear.

The photon is a quantum with this specific character: Successive emissions produce the same result as simultaneous emissions in the double-slit experiment. In quantum theory, a quantum with this specific character is called a Bose-Einstein quantum or a boson. The photon is a boson.

The electron is a quantum with a different character: Successive emissions produce a result different from that obtained by simultaneous emissions in the double-slit experiment. In quantum theory, a quantum with this specific character is called a Fermi-**Dirac** quantum or a fermion. The electron is a fermion.

There is a more elegant way to describe bosons and fermions, which will be explained later.

Section 5
Wave Mechanics and Quantum Mechanics

Erwin Schrödinger is one of the physicists whom the writers of this text revere in their hearts. Science requires more than sound experimental methods; extraordinary imagination and flashes of insight are also required to make rapid progress. We can feel the power of Schrödinger's gifted imagination in the scientific papers and other writings he left for posterity. He may be the scientist most gifted with incisive intuition. His profound manner of thinking enabled him to see through nature's veil of mystery. Let us now discuss Schrödinger's development of quantum theory as the fundamental framework of modern physics.

Einstein's doctrine of the light quantum, published in 1905, elucidated the meaning of the photoelectric effect, but complete verification of its theoretical implications did not occur until the discovery of the Compton effect in 1923, described as follows: The energy and momentum of a photon colliding with an electron become smaller because they are given in part to the electron at the moment of collision; as a result, the wavelength of the scattered light becomes longer. This transfer of energy and momentum is consistent with the **conservation laws** of energy and momentum for a collision of two particles. The Compton effect revealed that light, which had been thought of as a wave of the electromagnetic field, is a particle-like entity and that the collision process between particles must be taken into account when light interacts with fundamental constituents of matter, such as electrons. Young's interference experiment, on the other hand, provided evidence of the wave-like feature of light. Hence, the idea was established that light is a peculiar entity manifesting the features of both a wave of the electromagnetic field and a particle with energy and momentum; the way it appears depends on the method of observing it. Today any physical entity that may be considered as a wave of some field and also as a particle with energy and momentum is called a quantum, and a physical theory of quanta is called quantum theory. In this sense, Einstein's doctrine of the light quantum may be understood as a quantum theory of light.

Who on earth could imagine that not only light but also the fundamental constituents of matter, thus far believed without doubt to be particles, are all quanta?

In 1923 a young nobleman in Paris, **Louis de Broglie**, presented quite a bold hypothesis. He claimed that any fundamental constituent of matter is accompanied by a phase wave, that is, a peculiar wave different from an ordinary wave in which the amplitude changes. He suggested that this "phase wave" is a wave in which only the **phase** changes. Furthermore, the wave number and frequency of this phase wave were thought to be proportional to the momentum and energy, respectively, of the constituent of matter. That is, the momentum changes in proportion to the wave number, and the energy changes in proportion to the frequency. Each constant of proportionality was given by the **Planck constant**, which **Max Planck** considered to be the minimum amount of physical action in the sub-microscopic world.

De Broglie regarded this phase wave as a wave of the **matter field** and called it a **matter wave.** He advanced a new doctrine that every fundamental constituent of matter, such as an electron, was a quantum with both the wave nature of the matter field and a particle nature. Born at this moment was a quantum theory through which a fundamental understanding of matter and light could be sought.

As a matter of fact, ten years earlier (in 1913) a physicist in Copenhagen, Niels Bohr, had presented a new idea about the structure of the atom: the atom had been regarded by the chemist John Dalton in 1808 to be the fundamental constituent of matter without specifying its structure. Inside the atom, particles with a negative electric charge (i.e., electrons) go round and round a particle with a positive electric charge (i.e., the **nucleus**), just as the planets go around the sun in the solar system. However, there is an essential difference between the orbital motion of a planet and that of an electron. The planet is allowed to take any orbit around the sun, subject to classical mechanics as described by Newton, whereas the electron can take only restricted orbits around the nucleus, satisfying a specific property called Bohr's quantum condition. That is, there are a number of possible orbits for the electron circling the nucleus, and the electron keeps going around as long as it is in one of these orbits. This might explain the stability of the atom. Furthermore, Bohr postulated that the electron going around the nucleus can acquire or lose energy only by changing its orbit among those restricted orbits subject to Bohr's quantum condition. This idea was indispensable to explain the experimental fact that only light with restricted values of wavelength and frequency are emitted and absorbed by each atom. The energy differences between the restricted orbits subject to the quantum condition were postulated to be the same as the energies of light quanta emitted and absorbed by the atom. Recall that the energy of a light quantum or a photon was given by the frequency times the Planck constant in Einstein's quantum theory of light. If we go into technical detail, Bohr's quantum condition can be clearly stated in a somewhat mathematical expression:

> *Each electron moving around the nucleus can take only a restricted class of orbits, along which the integral of the momentum of the electron can take only limited values given by integral multiples of the Planck constant, that is, once the Planck constant, twice the Planck constant, three times, and so on.*

It was difficult for people to accept such an idea in 1913 when Bohr proposed the quantum condition, for it was foreign to the world view of classical physics, that is, Newtonian mechanics and **Maxwellian electromagnetism**. It was a total enigma why classical mechanics should suffer from such an incomprehensible constraint in the sub-microscopic world of atoms. There were no such theoretical constraints in the macroscopic natural world of classical mechanics. The idea of a matter wave, as proposed by de Broglie, was, indeed, the key to solving this mystery. If a phase wave is associated with an electron moving around the nucleus and its momentum equals the wave number of the phase wave times the Planck constant, then Bohr's quantum condition is nothing but the continuity of the phase wave. That is, the integration of its momentum along its orbit, coming full circle to any starting point around the nucleus, becomes the number of wavelengths, that is, the total number of crests or troughs of the phase wave around the nucleus. Then, Bohr's quantum condition yields that the total number of wavelengths of the phase wave, starting from a crest or trough and ending at the neighboring crest or trough, must be an integral number (one, two, three, and so on) but never a fractional number. In other words, the phase wave of an electron is tuned around the nucleus so that the phase wave is kept standing just like a tuning fork; it does not dissipate. An electron orbiting the nucleus is subject to a constraint (namely, Bohr's quantum condition) that is incomprehensible from the point of view of classical physics because it is a quantum acting as both a particle and a wave; such a thing does not exist in the framework of classical physics.

The concept of a matter wave, introduced by de Broglie, could not become established merely by explaining Bohr's quantum condition for atomic structure. If the idea of a matter wave were correct, an interference effect, like that obtained for light by Young, would be found in the electron beam simply because each electron in the beam would behave as a wave and suffer an interference effect. This is the very reason that the double-slit experiment for the electron was designed and performed. Of course, de Broglie, extending Einstein's doctrine of the light quantum and assuming the wave-particle duality not only for light but also for all fundamental constituents of matter, obtained a more universal doctrine of the quantum. It was considered a naive idea, one that de Broglie had obtained heuristically by analogy rather than developed as a concrete physical theory; therefore, it was rarely taken into a serious consideration by the scientific community. The only exception was Einstein, who told young physicists in Europe in the 1920s: "Read it. It is right, even though it looks strange." Among those young physicists was Erwin Schrödinger, who was at the University of Zürich in Switzerland at that time.

Schrödinger quickly realized that de Broglie's theory of the matter wave could not be comprehended and accepted by the scientific community because it lacked physical equations. That is, even though de Broglie had discovered the matter wave associated with the motion of an electron, he had simply supposed it to be a phase wave of the matter field with a wave number and frequency proportional to the momentum and energy, respectively, of the electron. To know how an electron moves in the sub-microscopic world, we need a physical law that prescribes the change of physical quantities such as momentum and energy. A law had been provided by Newton's law of motion or the equation of motion in classical mechanics. In classical mechanics, the equation of motion in Newton's law accounted for changes in physical quantities, but de Broglie could not find an alternative law in his theory of the matter wave. Without such a fundamental law of motion, de Broglie's idea of a matter wave could not predict the motion of an electron precisely. In fact, without the requisite fundamental law of motion, the theory that every fundamental constituent of matter is a quantum with both particle and wave natures was unsubstantiated.

To bring quantum theory to completion, a fundamental law of motion was required to show that the two important physical quantities for the motion of a quantum—momentum and energy—depend on the wave number and frequency of the matter wave. The wave number and the frequency of a wave designate how the wave changes in space and time, respectively. Therefore, the law of motion for a quantum would describe the way in which the quantum's associated matter wave changes in space and time through wave number and frequency.

Seeing the necessity for such a law, Schrödinger developed a new framework for quantum theory, named *wave mechanics*, in about two months while he was staying in Arosa, Switzerland. He gave the fundamental law of motion for a quantum in the form of a wave equation for the associated matter wave. Just as the orbital motion of a planet moving around the sun was understood by solving Newton's equation of motion, the motion of an electron orbiting the nucleus and the interference effect in the double-slit experiment for the electron were understood by solving the wave equation that Schrödinger discovered in 1926.

Chapter 2
Quantum Mechanics

Section 6
Schrödinger's Wave Equation

The **quantum theory** called wave mechanics, which **Erwin Schrödinger** completed within only two months, provided an epoch-making framework for mechanics that ingeniously included the wave-like aspect of **quanta**. No one knows today by what deep insight he discovered the wave equation that provided the fundamental law of motion for quanta such as **electrons**. He published a series of papers in 1926 that discussed a mathematical formulation of **quantum mechanics** in terms of the wave equation he had discovered (i.e., the Schrödinger equation). Schrödinger seemed most proud of his first paper. In many places in the text, we can see the amusement in his funny, smiling face. Indeed, the correct wave equation was derived by a unique means, a manipulation peculiar to Schrödinger. He did not use **de Broglie**'s relationships between **energy** and **frequency** and between **momentum** and **wave number** for a **matter wave** to arrive at his equation. After Schrödinger's second paper, however, an argument was set forth that made explicit use of de Broglie's insights into these relationships to enable everyone to understand the manipulation. The latter is much easier to follow. It was not until the 1980s that the scientific community understood the physical meaning of the mathematical manipulation peculiar to Schrödinger shown in his first paper (Yasue 1980; 1981).

Let us start by drawing a mental picture of de Broglie's matter wave, that is, a phase wave of the **matter field**. Suppose there exists an apparatus that looks like the face of a clock at each of the infinite points of space. Imagine these clocks at all points in space. Unlike an ordinary clock, not the hours but the angles are inscribed on the face, and, moreover, there is only one hand rotating clockwise. Although the hand becomes longer or shorter, we shall assume for the moment that its length is fixed. The length of the hand of this imaginary clock has no actual physical dimension of length, so it is useless to say that the hand is one meter or one foot long. Therefore, let us use the term "Erwin" to indicate an imaginary unit to measure the hand length of this imaginary clock, and let us suppose that each clock situated at each point in space has a hand one Erwin in length. Of course, it is only a clock-like apparatus; it has nothing to do with a clock that counts time. Therefore, we should give it a special name. Let's call this clock-like apparatus an Erwin meter. For awhile, the hand length of all the Erwin meters remains uniform, one Erwin long, but each hand may point in a different direction with a different angle in reference to a particular position, which, by way of analogy, would be 12 or zero hour on a regular clock. The angle changes from place to place and from time to time. We can think of the totality of an infinite number of Erwin meters as the matter wave proposed by de Broglie. It may seem strange, but there is no way to understand quantum theory without recourse to hard mathematical expressions. Only an elementary mathematical expression is needed to describe the angle.

A matter wave is nothing but a phase wave of the matter field associated with the motion of a quantum, which now may be well illustrated by comparing the tips of the hands (each one Erwin long) of all the Erwin meters at all points in space. If you draw a curve joining the tips of the hands, you will see a **wave** spreading over the whole of space, that is, a matter wave. Let us consider an electron moving with constant momentum and energy without disturbance from anything. Such an electron will be called a free electron. According to de Broglie, the matter wave manifests constant wave number and frequency in this case, and their values give the values of momentum and energy, respectively, if multiplied by the **Planck constant**. Here, the wave number shows how many times the hands of the Erwin meters turn over a spatial distance of one meter, and the frequency shows how many times the hand of an Erwin meter at a particular point turns during a time interval of one second. Therefore, the constancy of energy means that the hands of all the Erwin meters turn at the same pace as time passes; they are synchronized. Likewise, the constancy of momentum means that each Erwin meter manifests a constant increment of change in the spatial direction given by the direction of momentum. That is how we might illustrate de Broglie's idea of the matter wave.

Here let us recall a dynamical law pertaining to energy and momentum well known in **classical mechanics**. The energy of a free electron consists solely of kinetic energy since the electron is not disturbed. Kinetic energy is equal to **mass** times the square of **velocity** divided by two, and the momentum is given by mass times velocity.

$$[\textit{Kinetic Energy}]=(1/2)[\textit{Mass}][\textit{Velocity}]^2$$

$$[\textit{Momentum}]=[\textit{Mass}][\textit{Velocity}]$$

Physicists prefer to write down those equations more symbolically:

$$K = \frac{1}{2}mv^2$$

$$p = mv$$

Here, K stands for kinetic energy, m for mass, v for velocity, and p for momentum. Therefore, the energy of a free electron is nothing but the square of the momentum divided by twice the electron's mass.

[Energy]=[Kinetic Energy]=[Momentum]^2/2[Mass]

In symbolic notation,

$$E = K = \frac{p^2}{2m}$$

where E stands for energy. Although that equation was taken for granted in classical mechanics, only Schrödinger saw it as a fundamental physical law governing the change of the hand of each Erwin meter in space and time. What kind of physical law is it?

The momentum of an electron equals the wave number times the Planck constant in quantum theory.

[Momentum]=[Planck Constant][Wave Number]

Symbolically,

$$p = \hbar k$$

where k stands for wave number and $\hbar$ for the Planck constant. Then, the kinetic energy of a free electron becomes the square of the Planck constant times the square of the wave number divided by twice the electron's mass.

[Kinetic Energy]=[Planck Constant]^2[Wave Number]^2/2[Mass]

$$K = \frac{\hbar^2 k^2}{2m}$$

Energy is given by the frequency times the Planck constant in quantum theory, and it is equal to the kinetic energy obtained therein.

[Energy]=[Planck Constant][Frequency]
=[Planck Constant]^2[Wave Number]^2/2[Mass]

$$E = \hbar\omega = \frac{\hbar^2 k^2}{2m}$$

In symbolic notation ω denotes frequency.

This argument does not differ from the original idea of de Broglie. Hence, Schrödinger recognized that the relationship between the frequency and the momentum of a matter wave is a valid physical law governing the hand of each Erwin meter at each point in space. In other words, he found a local physical law that governs the spatiotemporal aspect of the fluctuation of a matter field spreading over the whole of space. Energy is given by frequency times the Planck constant, but if we consider energy in terms of each Erwin meter, it becomes equal to the rate of change in time (i.e., the **time derivative**) of the angle drawn by the hand times the Planck constant. Momentum is given by the wave number times the Planck constant, and it is equal to the rate of change in space (i.e., the **space derivative** or **gradient**) of the angle drawn by the hand of each Erwin meter times the Planck constant. As kinetic energy is the square of the momentum divided by twice the mass, it may be understood as the square of the gradient of the angle of each Erwin meter divided by twice the electron's mass. Energy consists of kinetic energy only in the case of a free electron, so the fundamental law for the motion of a quantum (e.g., an electron) discovered by Schrödinger is given in the following form:

All the Erwin meters distributed over the whole of space are related to each other in such a way that

[Planck Constant]{Time Derivative of}[Angle]
=([Planck Constant]^2/2[Mass])({Space Derivative of}[Angle])^2

holds.

In symbolic notation this equation becomes

$$\hbar\frac{\partial\theta}{\partial t} = \frac{\hbar^2}{2m}(\nabla\theta)^2$$

where θ stands for angle and ∇ denotes space derivative or gradient.

When we see the way the hand of every Erwin meter revolves subject to this fundamental law, we find there a form of the matter wave. Therefore, we see that the physical law discovered by Schrödinger governs the motion of a quantum in terms of the propagation of a wave of the matter field. This explains why this fundamental law is called Schrödinger's wave equation.

In the sub-microscopic world, the fundamental constituents of **matter** can no longer be considered simple point-like **particles** subject to **Newton**'s equation of motion, which claims that

> ***acceleration*** *times mass is equal to force.*
>
> *[Mass][Acceleration]=[Force]*

Or equivalently,

> *the second order time derivative of the position multiplied by mass is equal to force.*
>
> *[Mass]{Time Derivative of}^{2}[Position]=[Force]*

This equation is commonly referred to as Newton's second law. Symbolically we have

$$ma = F$$

and

$$m\frac{d^2x}{dt^2} = F$$

where a, F, and x stand for acceleration, force, and position, respectively. However, something is apparently missing, isn't it? Yes, it is. We have obtained a wave equation that only governs the matter wave associated with the motion of a free electron without any disturbance. What about the motion of a quantum subject to the action of a certain force? That is no problem at all. Although it may be difficult to consider only in terms of de Broglie's relationship between energy and momentum, it can be solved immediately once rewritten in the form of Schrödinger's wave equation.

In the case of the motion of matter subject to a force, classical mechanics asserts that the energy of that matter is given by the sum of its kinetic energy and the potential energy due to the action of the force. This potential energy represents the magnitude of the action on the matter in question, and its value at each place can change over time. In the case of the motion of the aforementioned free electron, the electron does not suffer from any such action, so its potential energy vanishes everywhere. This time the potential energy may take various values different from zero, and this fact will be taken into account in the form of Schrödinger's wave equation. That is, Schrödinger's wave equation offers an alternative to the dynamical law in classical mechanics that energy is given by kinetic energy plus potential energy. This is the alternative:

All the Erwin meters distributed over the whole of space are related to each other in such a way that

[Planck Constant]{Time Derivative of}[Angle]
=([Planck Constant]^2/2[Mass])({Space Derivative of}[Angle])^2
+[Potential Energy]

holds.

In symbolic notation,

$$\hbar\frac{\partial\theta}{\partial t}=\frac{\hbar^2}{2m}(\nabla\theta)^2+V$$

where V denotes potential energy.

This is Schrödinger's wave equation, which correctly describes the motion of an electron subject to the action of an external force from the point of view of the matter wave. When we pictured a matter field in terms of imaginary Erwin meters distributed over the whole of space, we assumed for the moment that the length of the hands of all the Erwin meters are fixed at unity; that is, they remain uniform, at least temporarily. As a matter of fact, this assumption is justified only when a quantum suffers from no disturbing action at all. To describe the motion of an electron subject to the action of a certain force, we have to think in general of the length of the hand of each Erwin meter changing from time to time and place to place. The hand of each Erwin meter at each point in space may change not only the direction in which it points but its length as well. Therefore, we have to take into account not only the change of angle but also that of length in the expressions of both time and space derivatives of the angle of the hand of each Erwin meter appearing in Schrödinger's wave equation.

There is an error in the form of Schrödinger's wave equation that we have obtained for an electron suffering from the action of a force: there is no room for the length of the hand of each Erwin meter. This may be due in part to the fact that we have used the well-known dynamical law in classical mechanics, which claims that energy equals kinetic energy plus potential energy. Although this law happened to remain valid in quantum theory for the motion of a free electron, it required modification for more general application. It was the physicist **David Bohm** who first proposed a way to modify the dynamical law in classical mechanics to obtain a correct form of Schrödinger's wave equation (Bohm 1952). Indeed, he found that the energy of a quantum is not given by the mere sum of kinetic energy and potential energy; he coined a new term—quantum potential—to represent something that must be added to the equation. Now energy equals kinetic energy plus potential energy and quantum potential. Bohm saw that the value of the quantum potential at each point depends only on the length of the hand of each Erwin meter located there and its gradient (i.e., space derivative). We can thank David Bohm for a correct form of Schrödinger's wave equation:

All the Erwin meters distributed over the whole of space are related to each other in such a way that

[Planck Constant]{Time Derivative of}[Angle]
=([Planck Constant]^2/2[Mass])({Space Derivative of}[Angle])^2
+[Potential Energy]+[Quantum Potential]

holds.

Symbolically, we havc

$$\hbar\frac{\partial\theta}{\partial t}=\frac{\hbar^2}{2m}(\nabla\theta)^2+V+Q$$

where Q stands for quantum potential.

A naive question may arise: What about the time derivative of the length of the hand of each Erwin meter? To see this, we need to introduce another consideration.

Section 7
The Erwin Meter and Probability

Schrödinger's wave equation is not the sole basis for quantum theory. It correctly describes the behavior of a matter wave in terms of a spatiotemporal pattern of the motion of the hand of each Erwin meter distributed over the whole of space; however, the matter wave is an insubstantial entity and can never be seen directly. When Schrödinger discovered wave mechanics in 1926, he seemed to have an idea of the physical meaning of the quantity measured and represented by the hand of each Erwin meter. That is, he saw the fundamental constituent of matter called a quantum, such as an electron, not as a point-like particle but instead as an infinite number of fragments of a quantum distributed over the whole of space in the sub-microscopic world. And he thought the fraction of the entire electron at any particular place was given by the length of the hand of an Erwin meter situated there.

However, even in the sub-microscopic world, the electrons we actually find are, indeed, point-like constituents of matter smaller than one trillionth of a centimeter in diameter, and we never observe any fragments of the electron. Then what on earth is the matter wave? How can something with such an insubstantial existence dominate the motion of matter?

While Schrödinger, de Broglie, and **Einstein** were confronting this challenging problem head-on and struggling to find the secret order behind wave mechanics or quantum theory, a convenient solution to their dilemma was proposed by the scientific community: "Gentlemen, forget it. Don't labor over such a useless issue." This was the pragmatic thinking: Don't inquire further into the physical meanings of the matter field and the matter wave. Just regard the squared length of the hand of each Erwin meter as nothing but an indicator of the **probability** of finding an electron at that place. Then, the hand of each Erwin meter situated at each point in space becomes nothing more than a measure aimed at the probable location of an electron, and the physical reality of the electron itself can be thought of as a point-like, fundamental constituent of matter. Max Born of Göttingen proposed that interpretation of the meaning of the hands of Erwin meters in 1926, and it was accepted by a significant number of influential people. Born's probabilistic interpretation, as it is called, has since become established.

Born's idea is, of course, merely a convenient interpretation of the matter wave. It provides us with nothing more than the possibility of correctly describing the motion of the fundamental constituents of matter in the sub-microscopic world. One can think of the whole theoretical framework of quantum mechanics as sufficient for the present by accepting Born's probabilistic interpretation. No one knows the answers to these questions:

- Is the motion of an electron, in fact, subject to the hand of each Erwin meter?
- Why does the hand's squared length give the probability of an electron being there?
- Why is the hand of each Erwin meter governed by Schrödinger's wave equation?

No one questions these things. Born's probabilistic interpretation seems to be accepted as common sense.

Did the scientific community lose interest in the cryptic mechanism of nature lying concealed behind Schrödinger's wave equation and in Born's probabilistic interpretation? No, not everyone. Throughout their lives, Erwin Schrödinger, who discovered wave mechanics, and Hideki Yukawa, a Nobel laureate for his discovery of meson theory, sought the yet unknown reality behind quantum theory. Yukawa looked for this reality in the discontinuous structure of space on the extremely fine sub-microscopic scale. He initiated a revolutionary school of thought in his *theory of elementary domains*.

Section 8
Elementary Domains, Reality and Our World

The first and second questions in the domain of basic research—"What is reality?" and "What is the world?"—are believed to have been answered by quantum theory and the **general theory of relativity**, respectively; however, we do not know why the fundamental assumptions or principles of those theories—that is, Schrödinger's wave equation, Born's probabilistic interpretation in quantum theory, and **Einstein's equation** in the **general theory of relativity**—are true. The assumptions or principles have been proven to be correct, but the scientific community does not know *why* they are valid. As we saw in the preceding section, what is relevant to our concern about current quantum theory with Born's probabilistic interpretation is the manner of dealing consistently with observable, dynamical variables and their statistical expectation values, not the reality of the world at the imperceptible sub-microscopic level. However, if we wish to put aside our conventional way of describing natural phenomena in order to achieve a true understanding of reality, we need to probe deeper for yet unknown physical objects behind the assumptions or principles of both quantum theory and the general theory of relativity.

In this connection it may be worthwhile to review the momentous lecture by Bernhard Riemann at Göttingen in 1854. Long before the advent of Einstein's general theory of relativity, Riemann had described precisely the fundamental notion of the geometry of our space (Riemann 1854). Namely, he revealed the possibility that our world is not a simple space in which Euclidean geometry remains valid; he suggested another notion of space in which the distance between two places is no longer fixed a priori. Riemann seems to have predicted Einstein's idea of general relativity in which distance in space has no a priori form but is attributed to the properties of the interaction of matter, as shown in Einstein's equation. As a matter of fact, Einstein used the mathematics of non-Euclidean geometry—specifically, Riemannian geometry, which originated from Riemann's 1854 lecture—to formulate his general theory of relativity.

There is no doubt about Riemann's brilliance and the scientific validity of his lecture, though most physicists still dismiss Riemann's real intention, as expressed in his lecture. He did more than pave the way for the general theory of relativity; he offered new insight into reality and our world. At the beginning of his lecture, Riemann called attention to the importance of physicists taking into account the possibility of a discrete geometry of space on the extremely sub-microscopic scale. Although he was very explicit in his message, Riemann's deep insight was ignored until more than one hundred years later when Yukawa proposed his atomistic world view, in which our world was thought to consist of an infinite number of discrete elements of space called *elementary domains* and reality was thought to be given by the transfer or exchange of energy among those domains (Yukawa 1966). Yukawa's idea of elementary domains is somewhat similar to the *atom* proposed by Democritus of Abdera and the *monad* suggested by Gottfried Leibniz; however, Yukawa has been most influenced by the world view of the Chinese poet Li Po enfolded in the following famous poem:

The natural world provides all things with temporary residences.

Time is a wayfarer for hundreds of generations traveling among temporary residences.

It is of great importance that Yukawa's theory of elementary domains provides a philosophical viewpoint from which to understand the two issues "What is reality?" and "What is the world?" and the cryptic order of nature behind the principles of quantum theory. In other words, the quantum mechanical laws governing the fundamental constituents of matter, described by the matter wave subject to Schrödinger's wave equation, can be derived naturally from the elementary domain structure of space on an extremely sub-microscopic scale. Yukawa's unique theory responds to these questions: "What is the world?" and "What is reality?" *"The world consists of an infinite number of discrete elements of space called elementary domains. ... Reality is given by the transfer or exchange of energy between those elementary domains"* (Yukawa 1966). According to the theory of elementary domains, any fundamental constituent of matter in the world is nothing but energy localized as and transferring between elementary domains; therefore, the fundamental property of the motion of matter can be reduced to that of the basic amount of energy passing from an elementary domain to the neighboring ones. Similar to the philosophical implications of Li Po's poem shown previously, the essence of the perceptible world is nothing but the ongoing transfer and relocalization of energy as elementary domains.

Let us compare Yukawa's answers to the two questions in the domain of basic research with the conventional answers in modern physics. The conventional view is that even on the sub-microscopic scale the world is spanned by empty space subject to the same continuum geometry perceptible on the macroscopic scale. Reality is given by the physical processes of **quanta** taking place in space. In other words, the conventional world view of modern physics does not force us to drastically change our concept of space, which provides the physical processes of quanta with a simple framework of Euclidean geometry on the sub-microscopic scale, though the concept of space changes on the extremely large scale, as shown by Einstein's general theory of relativity. That is, the perceptible continuum geometry of Euclid on the macroscopic scale must be reconciled with the general continuum geometry of Riemann on the extremely large scale, as on the cosmological scale.

Since the conventional concept of the perceptible continuum geometry of Euclid on the macroscopic scale has been maintained in formulating quantum mechanics, the fundamental law of motion for quanta is, hypothetically, a peculiar form described by Schrödinger's wave equation and Born's probabilistic interpretation. The reason that Schrödinger's wave equation and the probabilistic interpretation can provide us with the fundamental law of motion for quanta is not yet known. It may be supposed that the fundamental law of motion for quanta would be described in a completely different way than that given by Schrödinger's wave equation and the probabilistic interpretation if a totally different concept of space were put forth in developing quantum mechanics, which Yukawa did in his theory of elementary domains. Yukawa sought to derive the fundamental law of motion for quanta (in the peculiar form described by Schrödinger's wave equation and the probabilistic interpretation) from the discrete geometry of space on the sub-microscopic scale. We may find a complete parallelism between Yukawa's theory of elementary domains and Einstein's general theory of relativity: The latter aimed to derive the fundamental law of motion for matter (as described by Newton's laws of motion and universal gravitation in classical mechanics) from the general continuum (Riemannian) geometry of space on the extremely large scale. The former relies on the discrete geometry of space on the sub-microscopic scale to explain why the fundamental law of motion for quanta is given by Schrödinger's wave equation and the probabilistic interpretation.

In Yukawa's theory of elementary domains, the conventional concept of the perceptible continuum geometry of Euclid on the macroscopic scale can be obtained by a limiting procedure in which the dimension and separation of elementary domains vanish. Then, the world is described by the conventional continuum geometry of Euclid, in which the basic amount of energy localized within an elementary domain is regarded as a point-like, fundamental constituent of matter. It is this localized energy that can be regarded as a quantum in quantum theory.

Once the real nature of a quantum is found, in the continuum limit of the discrete geometry of space described by the theory of elementary domains, it is not so hard to obtain the fundamental law of motion for quanta. That is, it is not so difficult to support the assumptions or principles of quantum theory provided by Erwin meters, Schrödinger's wave equation, and Born's probabilistic interpretation. The hand of the Erwin meter located at each point in space is nothing but a physical quantity designating the average deviation of the distribution of elementary domains from the idealistic, uniform distribution. A quantum is the basic amount of energy residing in a single elementary domain and transferable to neighboring ones. Thus the motion of a quantum is necessarily subject to the deviation of the distribution of elementary domains. Note that the deviation consists of the average deviation, that is, the hand of each Erwin meter, plus the fluctuation around the average deviation. Schrödinger's wave equation is a different mathematical way of representing the most fundamental law in the theory of elementary domains governing the average deviation of the distribution of elementary domains by which localized energy is transferred among those domains. Born's probabilistic interpretation—namely, that the squared length of the hand of each Erwin meter stands for the probability of a quantum being at that point in space—turns out to be a statement of probability that can be derived naturally from the fact that the transference of energy is governed not only by the average deviation in the distribution of elementary domains but also by the fluctuation around the average deviation. Indeed, in the continuum limit of the theory of elementary domains, the transference of energy among elementary domains manifests randomness, due essentially to the fluctuation around the average deviation in the distribution of the elementary domains.

For those who are not satisfied with the conventional and provisional solutions to the issues "What is reality?" and "What is the world?" within the realm of quantum theory and the general theory of relativity, let us offer this alternative view: *The world consists of an infinite number of discrete elements of space called elementary domains, and Reality is given by the transfer or exchange of energy between those elementary domains.* We strongly recommend that you consider Yukawa's original thought.

Many physicists working in **quantum physics** still believe that space on the sub-microscopic scale is the same space that we perceive on the ordinary scale. Real things are seen as nothing more than quanta subject to the quantum mechanical law of motion. Although that is a naive image of the world and reality, the following essential issue has been generally ignored: There does not seem to be a "world" with "real things" on the quantum level, so how do quantum theory and quantum cosmology account for our experience of "world" and "reality"? Only a few physicists, including Yukawa, have made a serious attempt to answer that question by seeking an alternative to the mathematics of quantum mechanics based on the Euclidean concept of space.

Section 9
The Schrödinger Equation and the Wavefunction

Born's probabilistic interpretation claims that the squared length of the hand of each Erwin meter represents the probability of finding a quantum at each point in space. This hypothesis can be derived naturally from the extremely sub-microscopic (i.e., sub-nuclear) geometry of space in Yukawa's theory of elementary domains. What can be understood then about the nature of the hand of each Erwin meter? We posed this question in a preceding section, but we did not provide a correct physical law to describe how the length of the hand changes in time. Schrödinger's wave equation in the form we obtained by introducing Bohm's quantum potential is not yet complete. The equation could describe how the angle of the hand changes in time, but nothing was specified for its length.

As long as we accept the probabilistic interpretation derived naturally from Yukawa's theory of elementary domains, however, we can immediately get rid of this difficulty. Indeed, if the squared length of the hand of an Erwin meter turns out to be the probability of finding a quantum at that point in space, we can use a purely mathematical property satisfied by the concept of probability. Namely, the probability of finding a quantum at each point turns out to be unity (one) if summed up over the whole of space. The probability of finding an electron anywhere in the whole of space must equal one regardless of any other physical constraint, such as the fundamental law of motion described by Schrödinger's wave equation, simply because it should be somewhere in the whole of space. This fact necessarily imposes a mathematical constraint on the way that the probability of finding a quantum changes in time. This constraint is known as the **continuity of probability** or the **conservation of probability**, which can be described by the equation of continuity.

The time derivative of the probability of finding a quantum at each point in space compensates for the difference between incoming and outgoing probabilities represented by the ***divergence*** *of probability current.*

{Time Derivative of}[Probability]=
-{Divergence of}[Probability Current]

In symbolic notation,

$$\frac{\partial \rho}{\partial t} = -\nabla \cdot j$$

where ρ and j are probability and probability current, respectively, and $\nabla \cdot$ means to take divergence.

The left-hand side of this equation of continuity represents the time change of the probability of finding a quantum in a small spatial region surrounding each point in space. The right-hand side represents the growing rate of probability of finding a quantum in that small spatial region due to the fact that the probability itself flows as a probability current, just as the density of air is determined by the air current. Then, the equation of continuity claims essentially that there is neither a sink nor a source where probability leaves or enters. The probability cannot be increased here without causing a decrease there. It is conserved in the sense that the probability of finding a quantum somewhere in the whole of space must always remain equal to one. If it happens to be increased in a certain spatial region, then it should be decreased outside that region, so the total amount of probability over the whole of space is always kept in balance, like a seesaw.

The probability itself is given by the squared length of the hand of each Erwin meter. The probability current was found by Bohm to be the probability itself multiplied by the gradient (i.e., the space derivative or rate of change in space) of the angle of each Erwin meter and minus the Planck constant divided by the electron's mass. Then, the equation of continuity provides a supplementary equation to Schrödinger's wave equation in Bohm's form:

{Time Derivative of}[Hand Length]^2
=([Planck Constant]/[Mass]){Divergence of}
([Hand Length]^2{Space Derivative of}[Hand Angle])

This equation can be restated in more familiar language as follows:

The time derivative of the squared length of the hand of the Erwin meter at each point in space equals the divergence of the product of the squared hand length and the space derivative of the hand angle multiplied by the Planck constant and divided by the mass of the quantum.

Or symbolically,

$$\frac{\partial R^2}{\partial t} = \frac{\hbar}{m}\nabla \cdot (R^2 \nabla \theta)$$

where R stands for hand length. Furthermore, Bohm also found an explicit form of the quantum potential. It is given by

[Quantum Potential]
=-([Planck Constant]^2/2[Mass])
({Divergence of}{Space Derivative of}[Hand Length])/[Hand Length]

and in symbolic notation

$$Q = -\frac{\hbar^2}{2m}\frac{\nabla \cdot \nabla R}{R}$$

Consequently, Schrödinger's wave equation in Bohm's form, together with this continuity equation for probability in terms of the hand of each Erwin meter, provides a correct form of Schrödinger's wave equation.

All the Erwin meters distributed over the whole of space are related to each other in such a way that these equations hold:

[Planck Constant]{Time Derivative of}[Hand Angle]
=([Planck Constant]^2/2[Mass])({Space Derivative of}[Hand Angle])^2
+[Potential Energy]+[Quantum Potential]

{Time Derivative of}[Hand Length]^2
=([Planck Constant]/[Mass]){Divergence of}
([Hand Length]^2{Space Derivative of}[Hand Angle])

Symbolical expressions follow:

$$\hbar\frac{\partial\theta}{\partial t}=\frac{\hbar^2}{2m}(\nabla\theta)^2+V+Q$$

$$\frac{\partial R^2}{\partial t}=\frac{\hbar}{m}\nabla\cdot(R^2\nabla\theta)$$

Although this may be the correct form of the fundamental dynamical law of quantum mechanics, which describes the way in which the hand of each Erwin meter changes its angle and length from time to time and space to space, another clearer or simpler form would be needed to give the basic principle of quantum theory. Erwin Schrödinger was clever enough to think of a **Gauss plane** of complex numbers: that is, a **complex plane** placed upon each Erwin meter so that two real-number-valued quantities—the angle and the length of the hand of each Erwin meter—can be described by a single complex-number-valued quantity. The hand of each Erwin meter thus projected onto a Gauss plane can be regarded naturally as a complex-number-valued function of position and time, called a *wavefunction* and originally designated by Schrödinger as a Greek lower case "psi." Then, in terms of the wavefunction, the squared length of the hand of each Erwin meter becomes the **absolute square** of the wavefunction, and Schrödinger's wave equation in Bohm's form and the equation of continuity are combined into a single equation:

All the Erwin meters distributed over the whole space are related to each other in such a way that this equation holds:

[Imaginary Unit Number][Planck Constant]
{Time Derivative of}[Wavefunction]
=-([Planck Constant]^2/2[Mass])
{Divergence of}{Space Derivative of}[Wavefunction]
+[Potential Energy][Wavefunction]

In symbolic notation,

$$i\hbar\frac{\partial\psi}{\partial t}=-\frac{\hbar^2}{2m}\nabla\cdot\nabla\psi+V\psi$$

where ψ denotes wavefunction and i stands for **imaginary unit number**.

This equation can be restated more simply as follows: The time derivative of the wavefunction multiplied by the imaginary unit number and the Planck constant equals minus the divergence of the space derivative of the wavefunction times the squared Planck constant divided by twice the mass of the quantum plus the wavefunction multiplied by the potential energy.

It seems worthwhile to note here that Bohm's quantum potential disappears in this fundamental equation for the wavefunction, so the equation looks much more refined in this form. The explicit form of quantum potential admits a straightforward mathematical manipulation to cancel out with other terms. This is called the *Schrödinger equation*. Furthermore, the notion of the Erwin meter can also be concealed in the formulation of quantum theory, as is the case in the conventional framework of quantum mechanics. We obtain finally a complete, easy-to-handle form of the fundamental dynamical principle of quantum theory:

Every fundamental constituent of matter is a quantum with both particle-like and wave-like properties characterized by the Schrödinger equation in such a way that the probability of finding a quantum is given by the absolute square of the wavefunction.

[Probability]=|[Wavefunction]|^2

$$\rho = |\psi|^2$$

Section 10
The Double-Slit Experiment Revisited

A theory describing the motion of matter in the sub-microscopic world based on the Schrödinger equation and Born's probabilistic interpretation is called quantum theory or quantum mechanics. It has been verified by many physicists and supported by a large number of experimental results since Schrödinger presented its core framework in his series of papers in 1926. A typical experiment is the double-slit experiment, which we have already seen. Let us review this experiment to accustom ourselves to the way of thinking in quantum mechanics.

According to the probabilistic interpretation of quantum mechanics, an electron is a point-like, fundamental constituent of matter with almost vanishing dimension, and it manifests a spatial motion governed by the quantum theoretical law of motion in such a way that the probability of the electron being in each place is given by the squared length of the hand of the Erwin meter at that location. The hand of each Erwin meter changes in space and time subject to Schrödinger's wave equation. The image of a wave comes to mind once all the hands of Erwin meters are seen as a whole, though the wave is no longer called a matter wave in quantum mechanics. As we have seen, the range of the hands of Erwin meters is called the wavefunction. Because the absolute square of a wavefunction gives the probability, the wavefunction itself is also called the *probability amplitude*.

Let us imagine in the double-slit experiment that Erwin meters are placed at every point in space between the electron source situated on the far left and the detecting screen situated on the far right. Since the squared length of the hand of each Erwin meter stands for the probability of finding an electron at that point in space, all the hands would remain at vanishing length as long as no electron were emitted from the source. Once an electron is emitted from the source, then the probability of finding an electron in the vicinity of the source becomes greater, and the hands of the Erwin meters placed in the vicinity of the source become longer at that time. Of course, in quantum mechanics the lengthening of the hands of the Erwin meters outside the source represents the physical phenomenon that an electron is emitted from the source. Longer hands indicate a greater probability of finding an electron in the vicinity of the source. Although the length of the hand becomes longer only for those Erwin meters placed near the source, a spatial region of longer hands of Erwin meters transfers gradually to the right because the hands of neighboring Erwin meters are related to each other through Schrödinger's wave equation. Not only the length but also the angle of the hand changes in space and time subject to Schrödinger's wave equation. In this way, the length and the angle of the hands of Erwin meters in the vicinity of the electron source change one after another, moving to the right—that is, a kind of domino effect takes place until the change finally manifests in the region of the double-slit. As the experimental setting is symmetric up and down, the hands of the Erwin meters placed in the upper slit and those in the lower slit are the same in both length and angle.

Let us follow the change of the hands of the Erwin meters arrayed in space between the blackout plate with the two parallel slits and the screen. At the first instant that the hands of the Erwin meters placed in the slits become non-vanishing (i.e., non-zero) in length, all the hands of other Erwin meters between the blackout plate and the detecting screen would still be vanishing in length because the probability of finding an electron in the region between the blackout plate and the screen remains null. Therefore, the length of the hand of each Erwin meter vanishes everywhere on the detecting screen located to the right because the electron is still near the double-slit; it is neither in the region between the plate and the screen nor on the screen. There is no possibility of finding the electron on the screen at that time. We focus only on those Erwin meters placed on the screen. Although their hands are of vanishing length, the influence of non-vanishing hands in the region of the double-slit will propagate due to the "domino effect" and range over the screen as time passes. This is what the Schrödinger equation tells us. Recall that the length and the angle of the hands of Erwin meters in the vicinity of the electron source change one after another to the right until the change finally manifests in the region of the double-slit. However, since the experimental setting is symmetric up and down, the hands of the Erwin meters placed in the upper slit and those in the lower slit are the same in both length and angle. Therefore, the hands of the Erwin meters become non-vanishing in length in the regions of both the upper and lower slits when the change manifests in the region of the double-slit. The changes propagated from both slits give rise to changes in the length and angle of the hands of the Erwin meters on the screen.

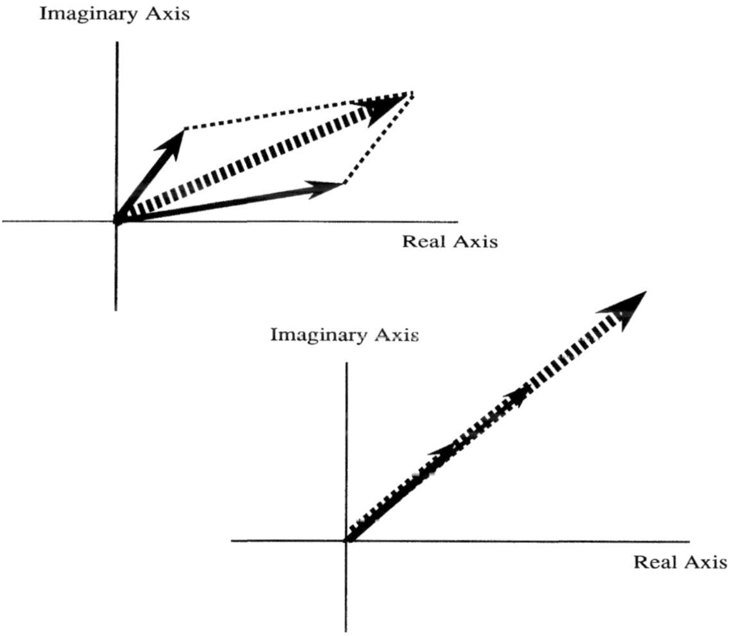

Figure 4. The parallel quadrangle law of addition of two complex numbers in the Gauss plane (top). Exceptional case of two complex numbers represented by two parallel arrows in the Gauss plane (bottom).

Here is an analogy that might help to clarify this phenomena: Imagine the still surface of a harbor protected by a sea wall with two adjacent, horizontal gates. When an ocean wave reaches the sea wall, it creates two overlapping waves perpendicular to the surface of the sea water. Then, the wave surging against the shore is given by the overlapping of the two waves. In this analogy, of course, the sea wall stands for the blackout plate, the two adjacent gates for the double-slit and the shore for the detecting screen. This movement is characteristic of the wave equation governing the motion of the interface between water and air. However, the waves consisting of the range of the hands of Erwin meters overlap in a peculiar way due to the specific form of the Schrödinger equation, which is nothing but a linear equation for the complex-number-valued wavefunction. That is, the overlapping of wavefunctions must be regarded as a sum of complex numbers subject to the parallel quadrangle law of addition in the Gauss plane. (See Figure 4.) In other words, the hand of each Erwin meter on the screen is given by overlapping the two hands propagated from the upper and lower slits in such a way that the resulting hand coincides with the diagonal of a parallel quadrangle drawn in the Gauss plane, whose two edge lines are determined by the two hands. In the exceptional case of two parallel hands, one cannot draw a normal quadrangle but can imagine an abnormal "quadrangle" crushed into a line with the same direction as the two hands and the length given by the simple sum of the length of the two hands. (See Figure 4.)

Let us imagine a number of Erwin meters on the screen. Among them are those given by overlapping two hands with approximately the same length and angle propagated from the two slits. Then, the length of the resulting hand, by means of the parallel quadrangle law of superposition, is found to be approximately the arithmetic sum of the lengths of the two hands. As the squared length of the Erwin meter represents the probability of an electron being in that place, there is a possibility of finding the electron there. It may happen for other Erwin meters on the screen that the two hands propagated from the two slits are of the same length but with opposite direction. The resulting hand determined by the parallel quadrangle law becomes vanishing in length, and the probability of an electron being there also vanishes. Thus, the possibility of finding an electron is greater at some places on the screen than at others. Furthermore, solving the Schrödinger equation reveals that places with a higher possibility and those with a lower possibility appear alternately on the detecting screen, resulting in the striped pattern. Namely, just as the wave surging on the shore (given by the overlapping of the two waves passing through the two gates of the sea wall) manifests an **interference** pattern, the lengths of the hands of the Erwin meters on the detecting screen also show an interference pattern. If electrons are emitted successively with a sufficient time interval between them to allow only one electron at a time to pass between the source and the screen, then it is most probable that the majority of electrons will hit the places where there is a greater possibility of finding an electron, so a striped pattern of electrons may be recorded on the screen. This is, indeed, a theoretical prediction of the result of the double-slit experiment for the electron.

The results of actual experiments—by Tonomura and others, for example—show complete agreement with the predictions of quantum mechanics (Tonomura et al. 1989). The distribution pattern of the squared length of the hand of each Erwin meter on the detecting screen coincides, in fact, with the striped interference pattern of the distribution of points on the screen detected in actual experiments. (See Figure 2.)

Section 11
Atomic Structure and the Stationary Wave

Quantum mechanics has revealed the electron's mechanism of motion in the sub-microscopic world. Louis de Broglie considered the electron to be a fundamental constituent of matter called a quantum with both particle-like and wave-like characters, and the wave-like character was attributed to the wave of probability amplitude given by the range of the hands of the Erwin meters. As we have seen, quantum mechanics has clearly explained the interference effect in the double-slit experiment for the electron.

De Broglie's original idea of the matter wave survives in the realm of quantum mechanics as the wavefunction or probability amplitude subject to the Schrödinger equation and the probabilistic interpretation of Born. In what way, then, can we understand the effect of Bohr's quantum condition on the electron's orbital motion around the nucleus in quantum mechanics, which de Broglie tentatively explained by applying the idea of the matter wave?

Let us look at the atomic structure within the realm of quantum mechanics. Since we are now accustomed to a quantum theoretical way of thinking, we will no longer use the intuitive image of Erwin meters but will rely on the elaborated notion of wavefunction. Although John Dalton proposed that the **atom** is the most fundamental constituent of matter in 1808, Ernest Rutherford of England found in 1911 an extremely small core with positive electric charge in the center of an atom by applying radiation to the atom. The core was called a nucleus. This discovery triggered the idea that the atom consists of electrons orbiting its nucleus like a miniaturized solar system in which planets go around the sun. In the actual solar system, the orbital motion is controlled by the gravitational attraction between the sun and the planets, whereas the motion of the electrons orbiting the nucleus was thought to be caused by the electric attraction of Coulomb force between the positive electric charge of the nucleus and the negative charge of electrons, due to the electromagnetic interaction described by

Maxwell's theory of electromagnetism. There is a problem, however, with electromagnetism in the motion of the electrons orbiting the nucleus. If we try to apply classical mechanics to this atomic model, the electrons in orbital motion would necessarily be subject to the attractive force so that the constant changes in their velocities would always confine the electrons near the nucleus; otherwise the atom would lose the electrons. In Maxwell's theory of electromagnetism, any charged particle, such as an electron, subject to a constant change in its velocity is known to lose energy by the constant emission of an **electromagnetic wave**. In **classical physics**, this effect is called *bremsstrahlung*, a German expression meaning "damping radiation." In classical mechanics, the more the electrons orbiting the nucleus lose their energy, the smaller their orbits become. The electrons are constantly drawn toward the nucleus as they lose their energy due to the *bremsstrahlung* effect. Thus, every atom is destined to be crushed completely by the constant contraction of the electrons' orbit around the nucleus, according to Newton's mechanical law as well as by the *bremsstrahlung effect,* as described by Maxwell's law of electromagnetism.

As we have already seen, in 1913 **Niels Bohr** proposed an incomprehensible constraint on the orbital motion of an electron around the nucleus; this constraint, called Bohr's quantum condition, achieved a breakthrough. Ten years later, de Broglie showed that Bohr's strange constraint was nothing but the physical condition of the matter wave of an electron; its physical condition being that of a standing wave around the nucleus. Of course, this idea assumes the existence of a matter wave behind the electron. Simply put, de Broglie proposed that each electron's orbit in an atom is a standing wave.

How then are we to understand the motion of an electron around the nucleus in quantum mechanics, in which de Broglie's naive notion of the matter wave is transformed into the wavefunction subject to the Schrödinger equation? The key concept is the condition of the standing wave. As the electron going around the nucleus is subject to the action of Coulomb's force due to electric charge, the electron has a potential energy, called the Coulomb potential, proportional to the product of the amounts of electric charge of the electron and the nucleus, and inversely proportional to the distance between them:

[Coulomb Potential]=[Electron Charge][Nucleus Charge]/[Distance]

Or symbolically,

$$V_C = \frac{ee_n}{r}$$

where V_C stands for Coulomb potential, e for electron charge, e_n for nucleus charge and r for distance. Therefore, the wavefunction determining the quantum mechanical motion of an electron changes in space and time according to the Schrödinger equation with the term for the potential energy replaced by this explicit form of the Coulomb potential. Such a change in the wavefunction around the nucleus is nothing but the behavior of a matter wave of an electron in the atom. If, as time passes, the wavefunction becomes vanishing in the spatial region surrounding the nucleus, then, according to the probabilistic interpretation, the electron no longer remains in the neighborhood of the nucleus, and the atom is stripped of electrons.

Hence, the probability of finding an electron would be vanishing in the spatial region far from the nucleus and would remain stationary in time everywhere so that the nucleus would always be enveloped by the non-vanishing probability of finding an electron. The wavefunction given by a complex number mathematically represents the hand of each Erwin meter; its **absolute value** (as a complex number) remains constant, but it turns around the origin of a Gauss plane with a constant rate. Such a wavefunction is said to be a stationary wavefunction. If we apply the condition of the stationary wavefunction of the electron orbiting the nucleus to the Schrödinger equation, we can reduce it to two separate equations, as follows:

The complex number giving the value of the wavefunction at each place around the nucleus simply turns around the origin of the Gauss plane with the absolute value remaining constant in time and the turning rate kept constant in time and uniform in space.

$$\psi = u(\cos \omega t - i \sin \omega t) = ue^{-i\omega t}$$

The time-independent, absolute value of the wavefunction manifests a spatial change such that this equation holds:

-([Planck Constant]²/2[Mass])
{Divergence of}{Space Derivative of}[Wavefunction]
+[Coulomb Potential][Wavefunction]
=[Planck Constant][Turning Rate][Wavefunction]

$$-\frac{\hbar^2}{2m}\nabla \cdot \nabla u + V_C u = \hbar \omega u$$

Here, in symbolic notation, u stands for the absolute value of wavefunction and ω for the turning rate.

The first equation claims the constancy of the electron's energy, based on these two conditions:

- The turning rate of the absolute value of the wavefunction around the origin of the Gauss plane is nothing but the frequency of the wave of probability amplitude (i.e., the matter wave).

- The frequency times the Planck constant equals the energy of a quantum.

This constant value for the energy appears in the second equation and ensures that the two equations can be combined to reproduce the Schrödinger equation.

The second equation describes how the absolute value of the wavefunction changes in space. As shown before, this equation requires the boundary condition that the wavefunction vanish in the region far from the nucleus. The second equation is called the stationary Schrödinger equation, and its unknown variables are the absolute value of the wavefunction and a constant designating the constant value of the energy of an electron moving around the nucleus. This type of equation is known as an eigenvalue equation or proper value equation, and there are many mathematical methods known to solve it. The stationary Schrödinger equation is considered an *eigenvalue equation*; therefore, the constant value of energy appearing there is often called the *energy eigenvalue*.

Let us call the time-independent, absolute value of the wavefunction simply a stationary wavefunction and take a brief look at one of the mathematical methods of solving the stationary Schrödinger equation for the simplest atom with a single electron orbiting the nucleus. This is the hydrogen atom. The stationary wavefunction is a real-number-valued function of position, and usually a Cartesian coordinate system is used to represent the position of the electron in space. However, taking into account the spatial symmetry around a nucleus placed at the origin of the Cartesian coordinate system, we have good reason to use the spherical polar coordinate system, as we do in the Earth's geography.

Suppose the nucleus to be an extremely small earth and the electron to be an extremely small satellite. A standard way to indicate the position of a satellite is to use three parameters, that is, longitude, latitude and altitude. The longitude is designated by the angle running from 0 to 2π in radians along the equator and measured from the intersection point of the Greenwich meridian line and the equator. The latitude is designated by the angle from $-\pi/2$ to $\pi/2$ in radians measured from the equator along the meridian line. The north and south poles correspond to the latitudes $\pi/2$ and $-\pi/2$, respectively. The altitude, of course, is a distance from the center or surface of the Earth to the satellite taking only a positive value. The center or surface of the Earth corresponds to the altitude zero. Similarly, we can follow the motion of an electron around the nucleus in terms of the longitude, the latitude and the altitude measured from the nucleus. The three **degrees of freedom** of the spatial displacement of an electron—that is, translations in the up and down, right and left, and front and back directions of the electron—can be replaced by two rotational degrees of freedom (latitude and longitude) plus a radial displacement degree of freedom (the altitude). Recall that the same description is frequently used to point out the position of a space shuttle, though in such a practical case the angle is expressed not in radians but in degrees. For example, the captain could tell us the actual position of his space shuttle by saying: "We just passed a point 35 degrees in latitude, 130 degrees in longitude, and 40 miles in altitude."

Since the stationary Schrödinger equation for the electron depends on the Coulomb potential, which is a function of the altitude only, the equation can again be broken down, this time into three separate equations. The first and second equations are the stationary Schrödinger equations describing the quantum theoretical dynamics of purely rotational degrees of freedom in latitude and longitude, respectively. The third equation is the stationary Schrödinger equation governing the quantum theoretical dynamics of a purely translational degree of freedom in altitude. The stationary Schrödinger equation for the rotational degree of freedom corresponding to longitude is not restricted to the case of a hydrogen atom; it is valid in any other case of atomic or molecular rotational dynamics.

The energy constant that appears in the original stationary Schrödinger equation is a sum of three components; the first one appears in the separate stationary Schrödinger equation for the rotational dynamics in latitude, the second one appears in the equation for the rotational dynamics in longitude, and the third one appears in the equation for the translational dynamics in altitude. That is, the energy of an electron moving around the nucleus is given by a component attributed to the rotation in latitude plus another component attributed to the rotation in longitude plus a third component attributed to the translation in altitude. The energy for the rotational degrees of freedom cannot take any value arbitrarily because the wavefunction must have the same value for two angles: 0 and 2π. Such a continuity constraint for the stationary wavefunction is found by a mathematical consideration, from which we derive this result: *any rotational energy in quantum mechanics can take only discrete values at even intervals.* This fact reflects the fundamental idea of the continuity of the matter wave in the atom proposed originally by de Broglie. As for the translational energy in altitude, the corresponding stationary Schrödinger equation can be solved by a standard method of mathematical analysis imposing the boundary condition that the wavefunction vanish for large values of altitude. The result is that the stationary wavefunction for the radial degree of freedom can be expressed by a function of altitude including one of the Laguerre polynomials. The translational energy is also found to take only discrete values, though not at even intervals this time.

The stationary wavefunction governing the quantum theoretical dynamics of an electron moving around the nucleus is obtained by combining the separate solutions for the rotational degrees of freedom in latitude and longitude with the solution for the translational degree of freedom in altitude. Correspondingly, the energy of an electron manifesting a quantum mechanical motion around the nucleus described by the stationary wavefunction is given by the sum of the three components, all of which take only certain discrete values, starting from the minimum value of each component and getting larger at intervals. This sum is the energy eigenvalue of the stationary wavefunction, and it also takes discrete values starting from a minimum value, as the components do. Corresponding to each discrete energy eigenvalue is a specific functional form for the stationary wavefunction. Those admissible values of the energy and the functional forms of the stationary wavefunction are called, respectively, the *energy eigenvalues* and the *energy eigenfunctions* in quantum mechanics.

It is extremely important to note that we must renounce the naive model of the atom described in terms of the planetary orbits of classical mechanics and Bohr's quantum condition when we investigate the actual motion of an electron around the nucleus within the realm of quantum mechanics. To describe the atomic structure correctly, we have to rely on the restricted group of energy eigenvalues and energy eigenfunctions derived from solutions to the Schrödinger equation. An electron subject to an energy eigenfunction is said to be in an energy eigenstate of the hydrogen atom. In general, a quantum described by an energy eigenfunction satisfying the stationary Schrödinger equation is said to be in the energy eigenstate corresponding to the energy eigenvalue. This energy eigenvalue determines the frequency of the wavefunction so that the energy of an electron equals the frequency times the Planck constant.

In conclusion, we can imagine an electron in quantum mechanics manifesting a stable configuration around the nucleus subject to the time-independent probability distribution given by the square of the stationary wavefunction of the energy eigenstate. This fact is not restricted to the case of the atom, of course. It applies to any other stable motion of the fundamental constituents of matter in the sub-microscopic world.

Chapter 3
Quantum Field Theory

Section 12
The Electromagnetic Field and Waves

Through the investigation of atomic structure, we have seen how **Bohr**'s quantum condition and **de Broglie**'s idea of the **matter wave** contributed to the birth of **quantum mechanics**. We have also seen why they had to be renounced and replaced by the Schrödinger equation and the wavefunction. Indeed, **classical physics**, failing to describe the stable configuration of **electrons** orbiting the **nucleus**, had to accept a strange law claiming that only orbits satisfying Bohr's quantum condition were realized as stable configurations of electrons around the nucleus. However, the meaning of Bohr's quantum condition was not at all clear in light of classical physics. A completely new system of fundamental laws of mechanics was needed to reveal why the atomic structure could manifest such a stable configuration of electrons orbiting the nucleus as was specified by Bohr's quantum condition. It was de Broglie who first explained the physical meaning of Bohr's quantum condition by assuming that an electron is a fundamental constituent of **matter** called a **quantum** equipped with a matter wave. Then, by elaborating de Broglie's idea of a matter wave, **Schrödinger** discovered a completely new system of fundamental laws of mechanics, that is, quantum mechanics. Although Bohr's quantum condition and de Broglie's matter wave contributed to the development of quantum mechanics, both of them were shown to have only figurative meaning and were replaced by the Schrödinger equation and the wavefunction. Those who now wish to go deeply into the terra incognita of **mind** and life with the help of **quantum theory** need to be familiar with more recent developments in this field. We need to understand the nature of matter in the sub-microscopic world, as revealed by quantum mechanics. To finally get a foothold in this strange world, we must revisit **Einstein**'s doctrine of the **light quantum** from the point of view of quantum mechanics.

Light was considered a **wave** of the **electromagnetic field** in classical physics. Einstein perceived it to be a particle-like entity in his attempt to explain the photoelectric effect. He called this **particle** of light a light quantum, though it is called a **photon** today. Einstein's theory of a light quantum demonstrated that the true nature of light cannot be adequately investigated within the realm of classical physics. Then, de Broglie extended Einstein's theory of the light quantum to matter and hypothesized that the fundamental constituent of matter, such as an electron, is a wave-like entity called a quantum. The matter wave conceived by de Broglie gave birth to quantum mechanics, though his original idea was later superseded by the wavefunction in quantum mechanics. Einstein's theory of the light quantum might undergo a similar transformation. Let's see how light is now regarded in the realm of quantum mechanics.

We became accustomed to a quantum theoretical way of thinking by investigating the motion of an electron in terms of the Schrödinger equation in quantum mechanics. In the case of the motion of an electron in the double-slit experiment or the hydrogen atom, it was sufficient to consider translational **degrees of freedom** in physical space, so we attributed an imaginary Erwin meter to each point in space. The Schrödinger equation explained the fundamental dynamical law governing the spatiotemporal change of the wavefunction, that is, a wave drawn in space by the range of the hands of Erwin meters. There, we could grasp the motion of an electron only by means of its probability distribution, which was represented by the spatial range of the squared lengths of the hands of Erwin meters.

Now, we proceed to the **quantum theory of light** by applying quantum mechanics to the electromagnetic field. The three translational degrees of freedom in the up and down, right and left, and front and back directions were sufficient to describe the motion of an electron in our three-dimensional physical space; however, the degrees of freedom for the electromagnetic field become rather complicated. The reason for the complication is that in the electromagnetic field as described by **Maxwell** in classical physics, there are an infinite number of invisible degrees of freedom. There are three invisible degrees of freedom per point of the three-dimensional physical space, and the total number of those invisible degrees of freedom becomes infinite because there are an infinite number of points in space. Therefore, we need a preliminary arrangement to illustrate the complicated degrees of freedom for the electromagnetic field, so we can understand the nature of the electromagnetic field without recourse to mathematical expressions.

First of all, let us call the three-dimensional physical space spanned by our universe simply space. Picture an imaginary, three-dimensional space superimposed like a ghost on each point in space. This ghostly, fictitious space will be called an electromagnetic space. The name is for the sake of convenience only; it has no deep significance. Another name, such as Maxwell space, would serve our purpose as well. A geometrical structure in which a kind of ghostly space adheres to each point in space is called a fiber bundle in mathematics, but simply a **field** in physics. We have just obtained such a geometrical structure in which an imaginary electromagnetic space is stuck on each point in space. This whole structure is the realm in which the electromagnetic field will be defined. (See Figure 5.)

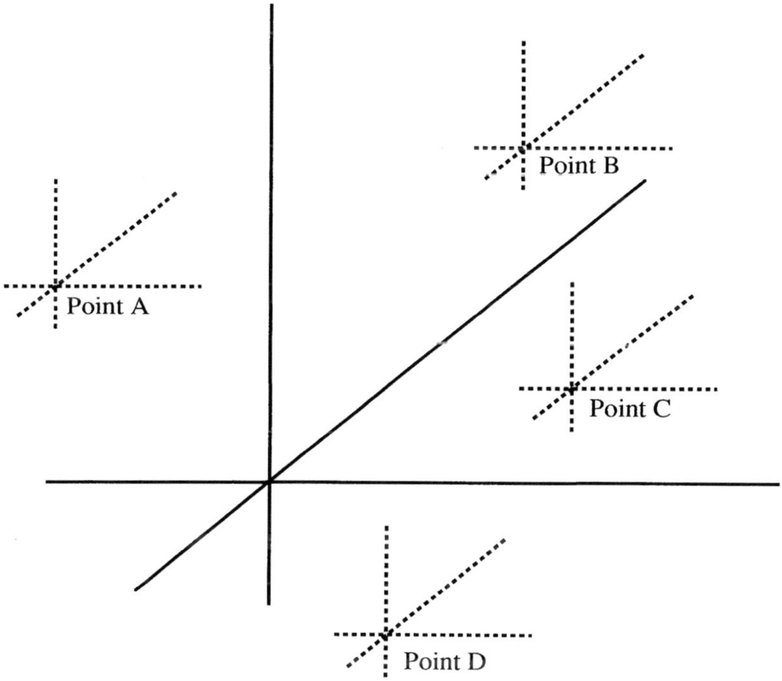

Figure 5. The electromagnetic field: An electromagnetic space (or Maxwell space) adheres in ghostly fashion to each point in space.

Let us imagine further a fictitious point moving in each electromagnetic space of the electromagnetic field; these imaginary points elucidate the degrees of freedom of the electromagnetic field. It is worthwhile to note that this is not a translational motion in real space but a ghostly motion in the fictitious electromagnetic space attributed to each point in space. Only by picturing the motion of the imaginary point in each electromagnetic space attributed to each point in space can we conceptualize the electromagnetic field. Now imagine an arrow pointing from the origin of each electromagnetic space to the moving fictitious point: Think of it as a three-dimensional **vector**. This vector will be called the *vector potential* of the electromagnetic field. (See Figure 6.) Then, the degrees of freedom of the electromagnetic field can be represented in terms of the vector potential as follows:

The vector potential can freely change the direction in which it points and its length in each three-dimensional electromagnetic space adhering to each point in space.

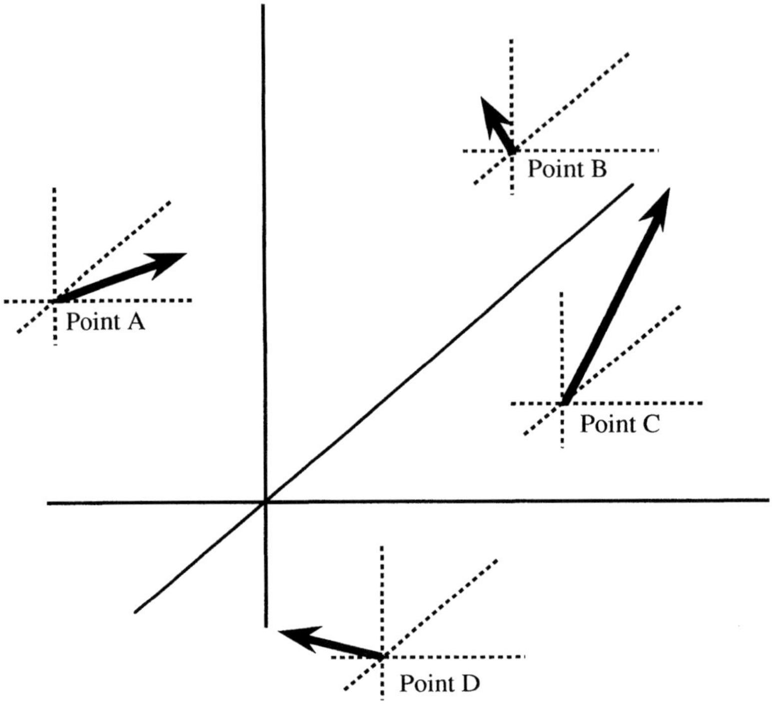

Figure 6. The vector potential of the electromagnetic field: The totality of vectors, that is, arrows, each of which is in an electromagnetic space.

In other words, the arrowhead of the vector potential manifests three translational degrees of freedom in each electromagnetic space. This may be the simplest way to picture the degrees of freedom of the electromagnetic field. It is often called the temporal gauge. Our investigation of the dynamics of the electromagnetic field within the realm of quantum theory is simplified if we represent the degrees of freedom of the electromagnetic field by means of the vector potential in the temporal gauge.

As we have seen, the degrees of freedom in the motion of an electron are simply up and down, right and left, and front and back translations in space, so it suffices to imagine an Erwin meter located at each point in space. Now, we have to consider the motion of the arrowhead of the vector potential in each electromagnetic space adhering in ghostly fashion to each point in space: its degrees of freedom appear to be up and down, right and left, and front and back translations in each three-dimensional electromagnetic space. Even though each electromagnetic space is a three-dimensional space, it differs from our physical space in that the electromagnetic space is an imaginary space we assumed simply for the purpose of illustrating the degrees of freedom of the electromagnetic field. There is no deep significance in designating three independent directions as "up and down," "right and left," and "front and back" in the electromagnetic space. We are using those terms for the time being simply as analogies to the familiar physical space in which we actually live. The Erwin meter is also meant to represent something that we cannot see. Remember that these imaginary meters depicted everywhere in space helped to clarify a quantum theoretical description of the motion of an electron. Now imagine an infinite number of Erwin meters occupying the points of the electromagnetic space that hovers like a ghost around each point in space. (See Figure 7.) The squared length of the hand of the Erwin meter situated at a point in the electromagnetic space represents the **probability** that the arrowhead of the vector potential exists at that point in the electromagnetic space.

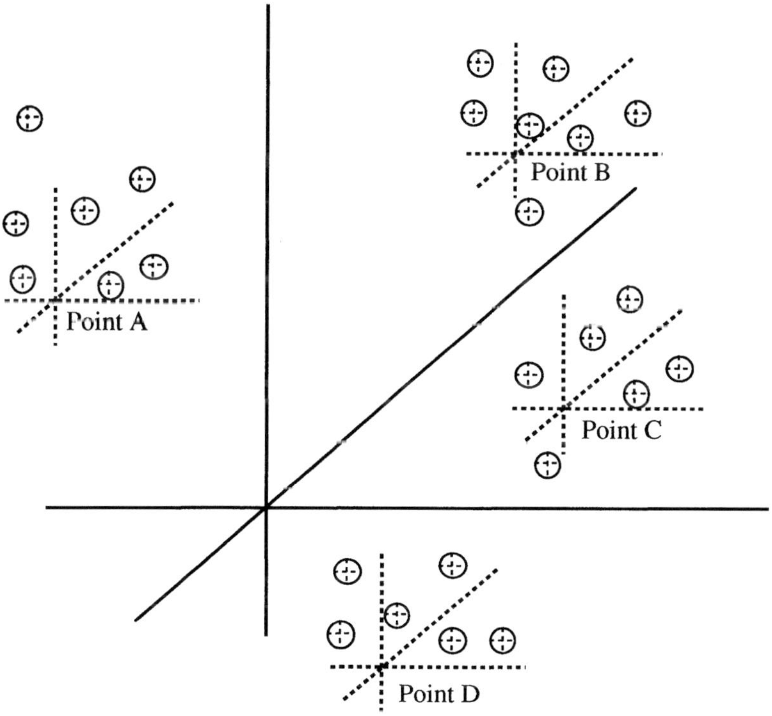

Figure 7. Erwin meters distributed throughout an electromagnetic space adhering in ghostly fashion to each point in space.

It may not be so difficult to describe the dynamics of the electromagnetic field in quantum theoretical terms if we imagine in this way an infinite number of Erwin meters distributed over the whole of each electromagnetic space. Once the reader has grasped the quantum theoretical description of the electron's movement around the nucleus—presented in terms of Erwin meters or, equivalently, of the wavefunction—the same method may help the reader to understand the motion of the arrowhead of the vector potential around the origin of each electromagnetic space. The difficulty here is that every point in space is burdened with an electromagnetic space; therefore, the motions of the vector potentials equal in number to all the points of space have to be taken into account simultaneously. In other words, we have to deal with an infinite number of wavefunctions and an infinite number of Schrödinger equations. The degrees of freedom of any field become infinitely large, and this fact makes the quantum theory of field dynamics much more complicated than quantum mechanics. The electromagnetic field is no exception by any means.

The dynamics of the vector potential in each electromagnetic space is described by **Maxwellian electromagnetism** in classical physics. There, the motion of an arrowhead of a vector potential in an electromagnetic space adhering to an arbitrary point in space is strongly related to the motion of the arrowheads of vector potentials in the electromagnetic spaces adhering to the neighboring points; therefore, the range of all the vector potentials over the whole space continuously changes in space and time, manifesting a wave form. This is nothing but a wave of the electromagnetic field, that is, an **electromagnetic wave**. The relationship among the vector potentials of neighboring electromagnetic spaces is given by Maxwell's wave equation as follows:

> *The arrowhead of the vector potential in each electromagnetic space adhering to each point in space changes in space and time so that this equation holds:*
>
> *(1/[Light Speed]^2){Time Derivative of}^{2}[Vector Potential]*
> *={Divergence of}{Space Derivative of}[Vector Potential]*

In symbolic notation, we have

$$\frac{1}{c^2}\frac{\partial^2 \mathbf{A}}{\partial t^2} = \nabla \cdot \nabla \mathbf{A}$$

in which c is a constant denoting the speed of light and $\mathbf{A}$ denotes the vector potential. The **divergence** of the **gradient** of the arrowhead of the vector potential stands for its second order rate of change in space as compared to that of the neighboring ones.

Maxwell's wave equation, which is an equation of motion for the electromagnetic field in classical physics, governs the motion of the vector potential in each electromagnetic space. However, in its present form, the equation of motion cannot be applied to the arrowhead of the vector potential moving around the origin in an electromagnetic space as it was applied to the electron moving around the nucleus because the arrowheads of vector potentials in neighboring electromagnetic spaces also appear in the equation of motion for the arrowhead of the vector potential in an electromagnetic space. Just a moment. We are looking for a way to understand Einstein's doctrine of the light quantum by applying the fundamental principle of quantum mechanics to the electromagnetic field. Einstein attempted to understand the true character of light in his theory of the light quantum by considering only monochromatic light. Now we are searching for the true character of light within the fully developed framework of quantum theory. Henceforth, it will suffice to consider an ideal monochromatic light, which turns out to be a special form of the electromagnetic wave advancing in a definite direction with a definite **wave number**. Now, let us investigate the dynamics of the electromagnetic field for this specific wave motion with definite wave number.

As shown previously, the wave number for the monochromatic wave equals the gradient (i.e., the first order rate of change in space) of the **phase** of the wave, so Maxwell's wave equation can be rewritten as follows:

> *The vector potential in each electromagnetic space adhering to each point in space points in a definite direction constantly and uniformly, and only its length changes in space and time in such a way that this equation holds:*
>
> *(1/[Light Speed]^2){Time Derivative of}^{2}[Vector Potential]*
> *=-[Wave Number]^2[Vector Potential]*

Symbolically, this becomes

$$\frac{1}{c^2}\frac{\partial^2 \mathbf{A}}{\partial t^2} = -k^2 \mathbf{A}$$

with k denoting wave number.

Vector potentials in neighboring electromagnetic spaces now vanish from the equation, so it can be thought of as an equation of motion in classical physics for the vector potential in each electromagnetic space. Furthermore, since only the wave motion of the electromagnetic field with a definite wave number is taken into account, the vector potentials in all the electromagnetic spaces apparently manifest the same behavior. In other words, the original, infinite number of degrees of freedom of the electromagnetic field in terms of the infinite number of translational degrees of freedom of the vector potentials in all the electromagnetic spaces adhering to all the points in space are now reduced to an extremely simple degree of freedom in which only the length of a single vector potential changes, with its direction fixed in a single representative electromagnetic space. The electron moving up and down, right and left, and forward and backward around the nucleus subject to Coulomb's electric attractive force appears much more sophisticated. So it is.

Section 13
Fields and Quanta

We proceed now to the quantum theory of light. As long as we were concerned with an ideal monochromatic light, the degrees of freedom of the electromagnetic field in classical physics were reduced to that of a single vector potential with a fixed direction but changing length along a single line in the representative electromagnetic space. The equation of motion for the arrowhead was obtained in the preceding section. Let us regard the arrowhead of the vector potential as a fictitious, point-like particle moving along a single fixed line passing through the origin of the representative electromagnetic space with its **mass** equal to unity. Then, the equation of motion for the arrowhead can be thought of formally as **Newton**'s equation of motion for this fictitious particle. The form of this equation of motion tells us that the particle, subject to an attractive force, may be understood to move along the single fixed line. The attractive force always pulls the particle toward the origin, and its strength is proportional to the distance between the particle and the origin. Such an imaginary particle is called in physics a one-dimensional harmonic oscillator. The harmonic oscillator suffers from a linear attractive force in the sense that the magnitude of the force is proportional to the distance between the harmonic oscillator and the origin of the representative electromagnetic space with a constant of proportionality given by the speed of light squared times the wave number squared; therefore, the potential energy of the harmonic oscillator is given by half of its squared distance from the origin multiplied by the speed of light squared and the wave number squared. This potential energy happens to be of the same form as the potential energy of an expanding and contracting spring, so we can picture a particle with its unit mass bound to the origin by a spring. (The origin here is a mathematical term indicating the origin of the Cartesian coordinate system of space, that is, a central point from which one measures the coordinates. In geometry the starting point is called the "origin.")

Finally, the easiest way to formulate the quantum theory of light is to use the quantum mechanics of the one-dimensional harmonic oscillator in the representative electromagnetic space. The same formulation as in the case of an electron suffices. We imagine an Erwin meter at each point on the straight line along which the harmonic oscillator moves. Then, as a convenient mathematical representation of the hand of each Erwin meter, we introduce the wavefunction taking a complex-number value at each point on the line. Recall that when we investigated the dynamics of an electron responsible for the atomic structure, Erwin meters were located at every point in space around the nucleus. The fundamental dynamical law in quantum mechanics claims that the wavefunction ranging over the whole of space changes subject to the Schrödinger equation. In the case of the harmonic oscillator, we need only consider Erwin meters at points on the line along which it moves. The Schrödinger equation can be split into the following two equations for the same reason that it was broken down for an electron moving around the nucleus:

Each complex number given as a value of the wavefunction at each point on the line along which the harmonic oscillator moves simply turns around the origin of each **Gauss plane** *with its absolute value kept constant in time and its turning rate kept constant in time and uniform on the line.*

The time-independent absolute value of the wavefunction manifests a spatial change along the line in such a way that this equation holds:

-([Planck Constant]^2/2){Space Derivative of}^{2}[Wavefunction]
-(1/2)[Light Speed]^2[Wave Number]^2
[Vector Potential]^2[Wavefunction]
=[Planck Constant][Turning Rate][Wavefunction]

In symbolic notation, this becomes

$$-\frac{\hbar^2}{2}\frac{d^2\Psi}{dA^2}-\frac{1}{2}c^2k^2A^2\Psi=\hbar\omega\Psi$$

in which Ψ stands for stationary wavefunction and A for the length of vector potential.

As we saw in the case of the hydrogen atom, the second equation is the stationary Schrödinger equation, and the two unknown variables—the constant turning rate and the time-independent length of the hand of each Erwin meter—can be obtained by solving the equation. Those variables are called the *energy eigenvalue* and the *energy eigenfunction* of the harmonic oscillator. Of course, both the translational degree of freedom and the potential energy are of much simpler forms than those of an electron moving around the nucleus, and the stationary Schrödinger equation can be quite easily solved as a simple eigenvalue equation.

As a result, the **energy** of a harmonic oscillator manifesting a quantum mechanical motion along the single straight line passing through the origin of the representative electromagnetic space described by the stationary wavefunction takes only certain discrete values, starting from the minimum value and getting larger at even intervals. Corresponding to each discrete energy eigenvalue is a specific functional form for the stationary wavefunction. Those admissible values of energy and functional forms of the stationary wavefunction are called, respectively, *energy eigenvalues* and *energy eigenfunctions* of the harmonic oscillator in the representative electromagnetic space. An arrowhead of the vector potential in the representative electromagnetic space manifesting a quantum theoretical dynamic subject to an energy eigenfunction is said to be in an energy eigenstate of the electromagnetic field. The energy eigenvalue determines the **frequency** of the wavefunction of the arrowhead so that the energy of the harmonic oscillator equals the frequency times the **Planck constant**.

The energy eigenvalue takes a restricted number of positive values at even intervals, and the smallest one equals one half the speed of light multiplied by the wave number and the Planck constant. To find the next to smallest value, we add the Planck constant multiplied by the speed of light and the wave number to the smallest value. The next larger value is obtained by adding again the same amount of energy, and so on. Each energy eigenvalue has a corresponding energy eigenfunction, from which we can easily obtain the probability distribution pattern of the harmonic oscillator along the line, though it may not be so important to know them explicitly. It seems more important to note the fact that the energy eigenvalue of the harmonic oscillator takes discrete values at even intervals.

Recall that the harmonic oscillator we have described within the framework of quantum mechanics in terms of the Schrödinger equation is nothing but the wave motion of the electromagnetic field corresponding to a monochromatic light with a definite wave number. What was regarded as a harmonic oscillator is a vector potential pointing in a definite direction in the representative electromagnetic space. The energy eigenvalue of the harmonic oscillator obtained within the framework of quantum mechanics is, therefore, nothing but the energy of the electromagnetic field of monochromatic light in quantum theory. In other words, the wave motion of the electromagnetic field corresponding to monochromatic light with a definite wave number in classical physics becomes subject to a quantum theoretical constraint in the sub-microscopic world: its energy can take only restricted values, spaced at even intervals equal to the Planck constant multiplied by the speed of light and the wave number, starting from a minimum value. What on earth does this fact tell us about the real nature of light?

These findings confirm Einstein's expectations about light. Light has a particle-like existence when it is measured in the form of energy. Its energy reveals the particle-like character of light. The Planck constant times the speed of light times the wave number equals the energy of this fundamental particle of light. The fundamental particle of light measured in terms of energy is called a light quantum or a photon. It is also known as a quantum of the electromagnetic field. The electromagnetic field with energy eigenvalues given by discrete values spaced at even intervals starting from the minimum value is understood to consist of zero photons, one photon, two photons, three photons, and so on. *This particle picture of light, arising from the quantum theoretical analysis of the wave motion of the electromagnetic field, supports the correctness of Einstein's doctrine of the light quantum.*

The minimum eigenvalue for the energy of the electromagnetic field is given by one half of the energy of a single photon, and it never vanishes. That is, the electromagnetic field bears one half of the energy of a single photon, even when there is no photon at all. This energy is called *zero-point energy* and is thought of as intrinsic energy due to *quantum fluctuations* in the *vacuum state. The vacuum state stands for a dynamical state of the electromagnetic field in which no photon exists, and it is nothing but the energy eigenstate corresponding to the minimum energy eigenvalue.* Other energy eigenstates of the dynamics of the electromagnetic field including one photon, two photons, three photons, and so on are called the single-photon state, the two-photon state, the three-photon state, and so on, respectively.

Of course, in Einstein's doctrine of the light quantum, the energy of a light quantum was given by the frequency times the Planck constant, and the **momentum** was given by the wave number times the Planck constant. There is no contradiction at all between Einstein's theory and the quantum theory of the electromagnetic field, which we have just discussed. The speed of light times the wave number equals the frequency. Then the Planck constant times the frequency becomes the Planck constant times the speed of light times the wave number and, thus, equals the energy of a light quantum. The momentum of the harmonic oscillator in quantum mechanics is given by the rate of spatial change (the gradient) of the hand of each Erwin meter along the line times the Planck constant. From a straightforward calculation of the rate of change of the wavefunction along the single fixed line in the representative electromagnetic space, taking into account the explicit form of energy eigenfunctions, the momentum of the electromagnetic field is also given by integral multiples of the fundamental value identical to the wave number times the Planck constant. A single photon is then thought to bear this amount of momentum.

The dynamics of the electromagnetic field with infinitely many degrees of freedom can be similarly described by quantum mechanics, and Einstein's picture of the light quantum can be replaced by the concept of a quantum or an energy quantum of the electromagnetic field. This prescription might no longer be restricted to the case of the electromagnetic field. If there exists a certain field with infinitely many degrees of freedom in the sub-microscopic world, one could apply quantum mechanics to investigate the wave motion of that field and perhaps, thereby, discover an energy quantum that has a particle-like existence. So thought Hideki Yukawa, who applied this fundamental theoretical framework of physics to fields other than the electromagnetic field.

Hideki Yukawa investigated the dynamics of the field of the **nuclear force** acting on nucleons, which are fundamental constituents of matter within the nucleus of an **atom**. (The **proton** and the **neutron** are both nucleons.) Yukawa used the same quantum theoretical framework to investigate this field as was applied to the electromagnetic field (Yukawa 1935). Yukawa predicted the existence of a new particle as an energy quantum of the field of the nuclear force, and, indeed, in 1937 Carl Anderson discovered that particle through experiments on **cosmic rays**. Today, this quantum of the field of the nuclear force is called the meson. Since then, the quantum theoretical framework of fields, that is, **quantum field theory**, has been shown to be the most fundamental physical law in the sub-microscopic world. Quantum field theory is required to explain not only various physical phenomena on the sub-microscopic scale but also those on the microscopic scale as well as the macroscopic scale, such as **superconductivity**. Our universe is no exception.

To investigate the way in which our universe was born, we have to know the dynamics of the universe itself when the universe was extremely small. Those fundamental dynamics are necessarily described by quantum theory. In Einstein's **general theory of relativity**, the dynamics of the universe is thought to be well illustrated by that of the field of the gravitational force, that is, the gravitational field in the universe. Therefore, quantum field theory applied to the gravitational field provides us with a completely new understanding of the birth of our universe. Such a domain of physics is called quantum cosmology, and it is now under active development.

Quantum field theory is nothing but the quantum theoretical laws of mechanics applied to the dynamics of any kind of field with an infinite number of degrees of freedom distributed over the whole of space on the sub-microscopic scale. Thus, quantum field theory covers the electromagnetic field, the field of the nuclear force, and the field of the gravitational force.

Section 14
Interaction between Matter and Light

Quantum field theory revealed that the dynamics of the electromagnetic field in the sub-microscopic world gives rise to the concept of the energy quantum corresponding to Einstein's notion of the light quantum or photon. According to quantum field theory, only a restricted class of dynamical states of the electromagnetic field with integral multiples of the fundamental energy quantum are realized, and they are regarded as representing integral numbers of photons, each of which is supposed to bear energy and momentum equal to the Planck constant times the frequency and the wave number, respectively, of the wave motion of the electromagnetic field.

What does that really mean? In quantum field theory, it is convenient to describe a fundamental physical process in terms of energy exchange. To explain this peculiar process, we will examine the interaction between matter and light within the realm of quantum theory. As we have seen, electrons, protons, and neutrons are all quanta, that is, fundamental constituents of matter that make up the atomic structure. In other words, matter is made up of many atoms in which electrons are in energy eigenstates, that is, manifesting specific quantum mechanical dynamics described by stationary wavefunctions around each nucleus, which is composed of protons and neutrons. Each electron in each energy eigenstate is understood to have energy equal to the corresponding energy eigenvalue.

Think of a single atom as a storehouse of energy in the form of electrons. It is also a storehouse of energy in the form of protons and neutrons confined to energy eigenstates inside the nucleus, but the amount of stored energy is so large that, when released, it destroys all the atomic structure of living matter, as in the case of a nuclear explosion. To investigate the basic issues of mind and **consciousness**, therefore, we need only consider the energy stored in the quantum theoretical dynamics of electrons. As we saw in the preceding sections, an electron moving around the nucleus can engage in only restricted dynamics as described by the energy eigenfunctions of the stationary Schrödinger equation. Correspondingly, the electron can have only restricted amounts of energy given by energy eigenvalues of the same equation. Thus, as a storehouse of the electrons' energy, a single atom can store only restricted amounts of energy by holding electrons in the corresponding energy eigenstates.

Depending on the number of electron-proton pairs constituting the atomic structure, the atom changes its character and is considered a different kind of atom. The atom with a single electron-proton pair is the hydrogen atom, which we investigated in the preceding sections. An atom with two such pairs is called the helium atom, with three such pairs the lithium atom, with four such pairs the beryllium atom, and so on. Even a hydrogen atom with a single electron can store more energy than another atom with many electrons if the electron remains in energy eigenstates with higher energy eigenvalues. *An electron is kept in energy eigenstates with higher energy eigenvalues by absorbing energy from the electromagnetic field, that is, by absorbing photons.* (A more detailed explanation is provided later in this text.)

Although a piece of matter may consist of so many atoms that their number is measured in terms of the **Avogadro number**, that is, about 10^{23}, a number of atoms seem to form a specific unit called a **molecule**. The quantum theoretical dynamics of electrons around the nuclei play an important role in the formation of a molecule. Moreover, the electrons' motion can be well described in the case of a molecule by the stationary Schrödinger equation with all the Coulomb forces of the nuclei taken into account; thus, the energy eigenvalues can be obtained. Thus, molecules can also be regarded as storehouses of energy in terms of electrons. However, they can store additional energy in terms of a larger number of degrees of freedom inherent in their geometrical forms. For example, any molecule with a geometrical configuration of atoms that lacks spherical symmetry manifests two rotational degrees of freedom in longitude and latitude, which can be described the same way as the orbital rotation of an electron around the nucleus of the hydrogen atom. One sees immediately then that such a molecule can store energies corresponding to the discrete energy eigenvalues of the purely rotational degrees of freedom.

Atoms and molecules also have translational degrees of freedom; they can move in space subject to a quantum mechanical law of motion. Their translational motion in space can be described by the stationary Schrödinger equation, taking their mutual interaction into account in the form of potential energies. The resulting energy eigenvalues are also limited to certain discrete values as long as they are trapped in a finite region by their mutual interaction.

Since the atoms constituting a molecule may change their geometrical configuration, the molecule manifests a number of small vibrations around the equilibrium configuration. Those vibrations, called molecular vibrations, can be thought of as a set of harmonic oscillators with different masses and spring constants, each of which admits discrete energy eigenvalues at evenly spaced intervals, as explained in the preceding sections. Thus, every molecule also stores restricted amounts of energy in the degrees of freedom of its molecular vibrations.

Matter made of as many atoms and molecules as the Avogadro number stores various but restricted amounts of energy corresponding to the many discrete energy eigenvalues for the rotational, vibrational, and translational degrees of freedom of the molecules. Each atom or molecule constituting matter also stores energy corresponding to the discrete energy eigenvalues of the quantum theoretical dynamics of electrons in the atom or molecule. These energy eigenvalues provide us with a systematic point of view from which to investigate the fundamental physical processes in the interaction between matter and light. Matter is capable of storing various amounts of energy but incapable of gathering or scattering energy by itself. To gather energy, it needs to receive energy from a physical entity different from matter, that is, from the electromagnetic field. To scatter energy, it needs to give energy to a different physical entity. The latter consists of photons, that is, energy quanta of the electromagnetic field spread over the whole of space. Note that only the electromagnetic field can be such a physical entity capable of exchanging energy directly with matter. Thus, *the fundamental physical interaction between matter and light is nothing but the exchange of energy between the energy eigenstates of matter and the several-photon state of the electromagnetic field.*

Let us suppose that the electromagnetic field manifests a wave motion with a definite frequency and wave number. Then, quantum field theory tells us, only a restricted value of the energy chosen from integral multiples of the fundamental amount can be carried by the wave motion, and this wave motion is understood as the quantum theoretical dynamics of the electromagnetic field containing several photons. Consider that the electromagnetic field contains a definite number of such photons, say several photons. That is, the electromagnetic field is in the several-photon state. The energy of each photon is given by the Planck constant multiplied by the frequency, and the photon's momentum is given by the Planck constant multiplied by the wave number. Since photons are nothing but energy quanta of the electromagnetic field enfolding all matter in our universe, it is quite plausible that energy quanta can be transferred freely between the electromagnetic field and matter, provided that the total energy is balanced. This last constraint is intuited from the ***conservation law** of energy*, which appears to be a fundamental principle of physics.

Thus, for an energy quantum of the electromagnetic field to be transferred to matter, there must be some degrees of freedom in matter that have energy eigenvalues whose difference is identical to the value of the energy quantum of the electromagnetic field. Since there are many different degrees of freedom in matter, such energy eigenvalues certainly exist. It follows then that an energy quantum of the electromagnetic field is received by matter, and we perceive this energy exchange to be the absorption of a photon by matter. Conversely, the degrees of freedom of matter give energy stored in one of their energy eigenstates to the quantum theoretical dynamics of the electromagnetic field as an energy quantum. This energy exchange is regarded as the emission of a photon from matter.

In quantum field theory, the interaction between matter and the electromagnetic field is described by energy exchange processes that lower the energy eigenvalue of matter and simultaneously raise that of the electromagnetic field and vice versa. The raising and the lowering of the energy eigenvalues of the electro-

magnetic field are perceived as the creation and the annihilation of a photon, respectively, by matter.

Section 15
The Matter Field

Einstein's doctrine of the light quantum has been well established in quantum field theory by regarding energy quanta of the electromagnetic field as fundamental constituents of light, that is, photons. Recall that de Broglie, inspired by Einstein's idea, envisioned a matter wave associated with the motion of each fundamental constituent of matter. A matter wave is a phase wave with a definite frequency and wave number, which represents the quantum theoretical motion of a fundamental constituent of matter, such as an electron, the energy and momentum of which are equal to the Planck constant times the frequency and the wave number, respectively. De Broglie perceived the **matter field** as filling the whole of space, just as the electromagnetic field is now seen to do. However, when the matter wave was replaced by the concept of the wavefunction in quantum mechanics, there seemed to be no place for the notion of a matter field. Does quantum field theory re-establish de Broglie's forsaken concept of the matter field?

In the preceding sections we saw that the principles of quantum field theory apply not only to the electromagnetic field but also to any other field with an infinite number of degrees of freedom. Can we then regard all the hands of the Erwin meters distributed over the whole of space as an infinite number of degrees of freedom of a certain field? Perhaps the wavefunction can be thought of as a field in which the space adhering in ghostly fashion to each point in physical space is the two-dimensional space of the Gauss plane, that is, a two-dimensional space in which each point represents a complex number (i.e., a **complex plane**). In the case of the electromagnetic field, a three-dimensional space, called the electromagnetic or Maxwell space, was thought to adhere in ghostly fashion to each point in physical space. This time, we will call each two-dimensional space of the Gauss plane an Erwin space. Let us apply the term "matter field" to a field in which an Erwin space is superimposed on each point in space. (See Figure 8.) The dynamics of the matter field is exhibited, of course, by Schrödinger's wave equation, just as the dynamics of the electromagnetic field is described by Maxwell's wave equation. Here we can apply quantum field theory to the matter field just as it was used to investigate the dynamics of the electromagnetic field.

There is an apparent distinction, however, between the matter field and the electromagnetic field, which calls for an adjustment in our description of the two. The matter field carries only one fundamental constituent of matter, that is, only one quantum, whereas the electromagnetic field carries an unrestricted number of photons. The matter field is essentially nothing but a wavefunction governing the motion of a quantum in the sub-microscopic world. Such a restriction will give rise to a fundamental constraint in developing a quantum theory of the matter field. The following examines that constraint.

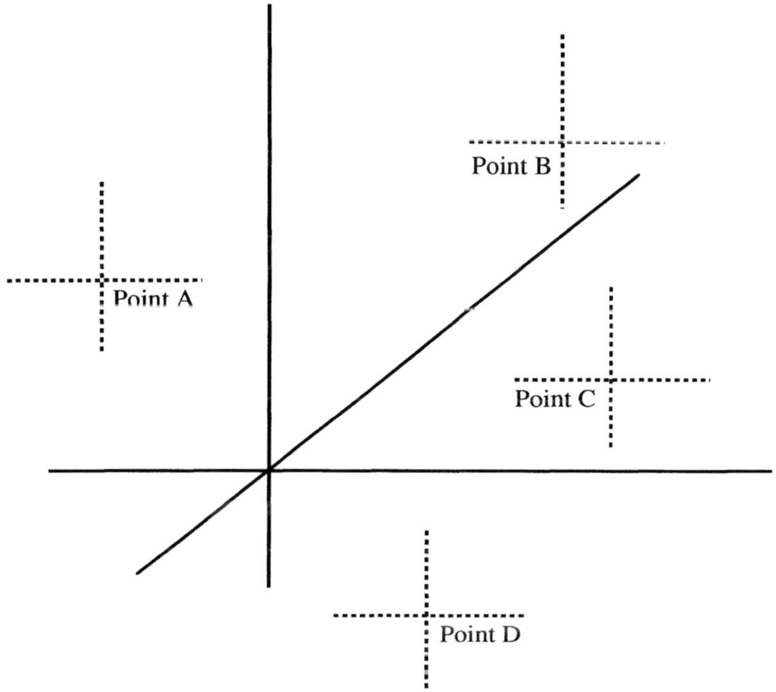

Figure 8. The matter field: An Erwin space (or a Gauss plane) adheres in ghostly fashion to each point in space.

We shall consider the dynamics of electrons constituting matter that interacts with the electromagnetic field. Recall that the Schrödinger equation tells us that there exist numerous specific, dynamical states characterized by stationary wavefunctions. They are energy eigenstates described by energy eigenfunctions and energy eigenvalues. In quantum mechanics, each energy eigenfunction governs the dynamics of an electron. In quantum field theory, we are going to present a somewhat different point of view, one in which wavefunctions are regarded as wavc motions of the matter field of the electron. Hence, each energy eigenfunction or stationary wavefunction turns out to be a stationary or standing wave of the matter field. An advancing wave motion of the electromagnetic field appears to carry energy when a number of energy quanta, that is, photons, exist there. An advancing wave motion of the electromagnetic field may represent no photon at all if it is in the vacuum state, that is, the no-photon state. If it is in a several-photon state, the same advancing motion of the electromagnetic field represents several photons.

A standing wave of the matter field is also thought to bear energy equal to the corresponding energy eigenvalue when an electron exists in the corresponding energy eigenstate. In other words, we drastically change our perception of electrons or any other quanta: they are nothing but energy quanta of the matter field. *The fundamental constituents of matter, such as electrons, neutrons and protons, are now seen as the energy quanta of the matter field.* However, there is a clear difference between the matter field and the electromagnetic field: each advancing wave motion of the electromagnetic field can bear any number of energy quanta—that is, no photon, one photon, two photons, three photons, and so on—but each standing wave of the matter field can bear at most only one energy quantum, that is, no electron or one electron. Without acknowledging this distinction between the wave motions in the two fields, we cannot apply quantum field theory correctly to the investigation of the fundamental processes of matter, nor can we show the equivalence of quantum field theory to quantum mechanics in which each wavefunction is thought to describe the dynamics of a single electron.

Quantum field theory provides a unified framework that describes the fundamental processes of matter and light. The essential difference between matter and light is also clarified therein. Namely, each advancing wave motion of the electromagnetic field may have any number of energy quanta, whereas each standing wave of the matter field can have at most only one energy quantum. In quantum field theory, therefore, two different kinds of fields are taken into account. The first kind of field is characterized by a wave motion that bears any number of energy quanta; this is called the Bose-Einstein field or Bose field. The second type of field is characterized by a wave motion that bears, at most, one energy quantum; this is called the Fermi-Dirac field or Fermi field. An energy quantum of a Bose-Einstein field is called a Bose-Einstein quantum; that of a Fermi-Dirac field is referred to as a Fermi-Dirac quantum. A Bose-Einstein quantum is also called simply a boson, and a Fermi-Dirac quantum a fermion. The electromagnetic field is a Bose-Einstein field and photons are bosons, whereas the matter field of the electron is a Fermi-Dirac field and electrons are fermions. Protons and neutrons are also fermions, though they are not relevant to basic research pertaining to life and consciousness.

It seems worthwhile to recall that we have introduced the notions of Bose-Einstein and Fermi-Dirac quanta in relation to the double-slit experiment. A Bose-Einstein quantum was described as a quantum with this specific character: in the double-slit experiment, successive emissions of such quanta produce the same result as simultaneous emissions. A Fermi-Dirac quantum, on the other hand, produces a different result if it is emitted successively or simultaneously in that experiment. Such naive characterizations of bosons and fermions are supplanted by more elegant descriptions here within the realm of quantum field theory: Since photons are bosons, any number of photons can occupy the advancing wave motion of the electromagnetic field. Thus, the wave motion of the electromagnetic field itself in the double-slit experiment cannot be modified by the actual number of photons in the wave motion. In other words, successive emissions of photons, in which only a single photon occupies the advancing wave motion, produce the same result as simultaneous emissions of photons, in which many photons occupy the same advancing wave motion.

Since electrons are fermions, only a single electron can occupy the advancing wave motion of the matter field of an electron. Thus, the wave motion of the matter field itself in the double-slit experiment is modified by the actual number of electrons in the wave motion. In other words, successive emissions of electrons, in which only a single electron occupies the advancing wave motion, produce a result totally different from that of simultaneous emissions of electrons. In the case of simultaneous emissions of electrons, all the electrons cannot occupy the same advancing wave motion. Each of them occupies a separate advancing wave motion of the matter field different from all the others, so the resulting **interference** pattern is different from that of successive emissions of electrons.

Section 16
Disorder and Order

Our world is made of matter and light. The fundamental processes taking place between them fall within the scope of quantum field theory. Within that framework, every constituent of matter or light is regarded as an energy quantum associated with a certain wave motion of the matter field or the electromagnetic field, respectively; thus each manifests both a particle and a wave nature. The interaction between matter and light is understood by means of energy exchange processes between energy eigenstates of the matter field and several-photon states of the electromagnetic field. Raising and lowering energy eigenvalues of the electromagnetic field are regarded, respectively, as the creation and the annihilation of photons.

We may encounter some difficulty, however, when we attempt to solve all the fundamental equations involved in the quantum field theoretical description. First of all, we have at least as many fundamental equations as the degrees of freedom of matter, which exceeds the Avogadro number. The situation seems worse for those of the electromagnetic field which are, indeed, infinite. The second problem is that the Schrödinger equation, serving as the fundamental equation, can be solved explicitly only for simpler dynamical systems with a few degrees of freedom, such as the harmonic oscillator and the hydrogen atom. Of course, this does not imply incompleteness of quantum field theory. Rather, our problem is that a proper mathematical method for solving the fundamental equations of quantum field theory has not been found yet.

The lack of such a method has forced physicists to find appropriate approximations so that certain physical aspects of matter and light might be understood qualitatively, and the fundamental equations of quantum field theory might be reduced to more manageable phenomenological or heuristic equations. For example, all types of matter in our world share a common physical aspect described phenomenologically by the thermodynamics of Nicolas Carnot and Rudolf Clausius in classical physics. Such a description, which does not pertain to the detailed, sub-microscopic structure of matter, maintains that matter is in the state of thermal equilibrium, or simply in thermal equilibrium, if the totality of energy exchanged between all the macroscopic parts of matter and the electromagnetic field balances. Furthermore, a systematic approximation capable of describing or analyzing complex physical systems of the fundamental constituents of matter and light in thermal equilibrium has been obtained. It is called quantum statistical mechanics or equilibrium quantum statistical mechanics. Most of the theoretical results concerning the physical investigation of matter have been obtained within the realm of quantum statistical mechanics.

It seems important here to note the significance of quantum statistical mechanics. The basic intention of quantum statistical mechanics is to incorporate **quantum physics** into a theoretical description of the macroscopic features of matter in thermal equilibrium by introducing reasoned approximations to the fundamental principles. The approximation assumes that the typical nature of matter in thermal equilibrium does not change significantly even if the mutual correlation between the dynamics of the fundamental constituents of matter is neglected. That is, each energy eigenstate may absorb or emit an energy quantum *independently* of the absorption or emission of energy quanta by other energy eigenstates, provided that the total energy exchange between all the macroscopic parts of matter and the electromagnetic field balances. In such a case, the dynamics of the fundamental constituents of matter are considered to be uncorrelated. Notice that each energy eigenstate of actual matter not in thermal equilibrium absorbs or emits energy quanta in a way highly dependent upon the absorption or emission of energy quanta by other energy eigenstates, as described by a large number of Schrödinger equations.

Quantum statistical mechanics has had much success in explaining typical physical aspects of matter. This fact indicates that the assumed absence of mutual correlation in the actual dynamics of matter that is composed of as many ingredients as the Avogadro number adequately approximates these features. Picture each standing wave of the matter field receiving or releasing an energy quantum by absorbing or emitting photons regardless of what the other standing waves do. Such uncorrelated physical motion in the dynamics of many quanta may be regarded as disordered dynamics, and matter in this case is said to manifest thermal disorder. The heat of matter is nothing but a measure of the average number of energy quanta borne by each standing wave of the matter field when the matter is in thermal equilibrium and manifests thermal disorder. Thus, quantum statistical mechanics provides a good approximate description of the macroscopic features of matter as long as the dynamics of the fundamental constituents of matter manifests thermal disorder.

Due to this success, quantum statistical mechanics has been expanded somewhat to include matter nearly, but not quite, in a state of thermal equilibrium. With this development, the chemical and biochemical reactions of atoms and molecules have fallen within the scope of the modern analytical method of investigation, giving rise to **molecular biology**. The medical and biological sciences of today, aided by the theory of quantum statistical mechanics, provide a detailed microscopic understanding of biochemical processes taking place in living matter. It is not usually mentioned explicitly that quantum statistical mechanics is merely an approximative framework, limited in its validity to the energy exchange dynamics of quanta manifesting thermal disorder without mutual correlation. An unfortunate result of this limited understanding of quantum statistical mechanics is that some persons in the medical and biological sciences expect quantum statistical mechanics to be reliable in the quantum theoretical investigation of living matter. However, the dynamics of quanta involved in the fundamental processes of living matter do not manifest thermal disorder at all; they exhibit order, as Erwin Schrödinger first pointed out (Schrödinger 1944). Matter is said to be in the state of order if the dynamics of each energy eigenstate manifests strong mutual correlation with other energy eigenstates. Picture each standing wave of the matter field receiving or releasing an energy quantum by absorbing or emitting photons in perfect accordance with the energy transformations in other standing waves.

When the famous mathematician **Johann von Neumann** saw scientists investigating the dynamics of water by using the **Euler equation** for a perfect fluid without viscosity, he laughed at the dynamics of "dry" water. Water is known to be a typical fluid with high viscosity, so the dynamics of water can hardly be described by the Euler equation, which describes the dynamics of non-viscous fluid. Similarly, we should keep in mind that investigating living matter through molecular biology and biochemistry based on quantum statistical mechanics may not render a true picture; that is, we might be looking for "dry" living matter. Living matter manifests not disorder but order or strong mutual correlation in its dynamics. Living matter differs from non-living matter drastically in that it decreases **entropy** and disorder and increases **negentropy** and order, even when the supply of energy comes in a completely disordered form. Non-living matter in thermal equilibrium is always kept at maximum entropy or disorder. Quantum statistical mechanics has provided a good approximate description of non-living matter. Another approximate framework derived from quantum field theory—opposite, in a sense, to quantum statistical mechanics—is required to investigate living matter. The order or strong mutual correlation in the energy exchange dynamics of living matter demands that we look at those processes from a completely new standpoint within quantum field theory, a standpoint quite unlike the one through which we examine the disordered dynamics of the fundamental constituents of matter in thermal equilibrium.

Section 17
The Collective Mode and the Macroscopic Wavefunction

As long as we are concerned with non-living matter in thermal equilibrium, quantum statistical mechanics may provide a good approximation to quantum field theory. However, this will not be the case when our investigation takes us into the terra incognita of the two remaining basic fields of research, that is, life and consciousness. There, contrary to what was observed in the case of the disordered dynamics of the fundamental constituents of non-living matter in thermal equilibrium, order and strong mutual correlation must be incorporated into the approximation of the fundamental principles. The reason is that living matter is nothing but matter with these peculiar features, as noted earlier: it decreases entropy and disorder and increases negentropy and order, even when the supply of energy comes in a completely disordered form. These features were originally pointed out by Schrödinger (1944). His point of view, however, was underestimated until Hiroomi Umezawa emphasized its importance in the early 1960s.

Umezawa emphasized the importance of incorporating the notion of the collective mode into the investigation of the fundamental processes of living matter so that order and strong mutual correlation in the dynamics of the fundamental constituents might be well described by quantum field theoretical notions. In the year 1967, Umezawa and Luigi Ricciardi published a momentous paper that focused on the highly systematized functioning of brain tissue as the most elaborate example of the typical aspects of living matter (Ricciardi and Umezawa 1967). Just as thermal equilibrium is the key concept in approximative physical analyses of non-living matter, the collective mode is of central importance in the study of living matter. Therefore, for the purpose of developing a quantum theory of consciousness in the third part of the present volume, we should become familiar with the quantum field theoretical framework of the collective mode. This will be the final topic in the first part of this volume, which so far has been an exposition of a "person-to-person" instruction in the basics of quantum mechanics and quantum field theory.

Let us start by thinking about order in the dynamics of energy quanta of the matter field. Matter in our world consists of numerous atoms and molecules, measured in terms of the Avogadro number, each of which has several energy eigenstates for the dynamical degrees of freedom of electrons as well as certain molecular conformational degrees of freedom, such as molecular vibration and rotation. Those energy eigenstates, which are identical for all the atoms or molecules of the same kind, are called elementary energy eigenstates. For example, the elementary energy eigenstates of all hydrogen atoms are identical, and the elementary energy eigenstates of all oxygen molecules are identical. In other words, from the point of view of quantum field theory, matter is perceived to consist of multitudes of identical groups of standing waves corresponding to the elementary energy eigenstates. That is, the standing waves of elementary eigenstates can exist everywhere in the matter field, and their total number on average remains constant. Quantum field theory, therefore, can describe the dynamics of the fundamental constituents of matter as energy quanta occupying the standing waves of elementary eigenstates everywhere in the matter field. The dynamics of the fundamental constituents of matter viewed in this way is called the occupation dynamics of energy quanta of the matter field. If a standing wave of the matter field corresponding to an energy eigenstate happens to bear an energy quantum, it is said to be occupied by a quantum. The number of quanta occupying each standing wave or energy eigenstate is unlimited for bosons but limited to one for fermions. The dynamical degrees of freedom of electrons and the conformational degrees of freedom of molecules give rise to fermions and bosons, respectively.

It is important to understand that the occupation dynamics of the energy quanta of the matter field can be described by following the change in time of the occupation number of each standing wave or energy eigenstate. It is customary in quantum field theory to think of an energy quantum as being *created* when the occupation number of a certain wave increases by one. Similarly, an energy quantum is thought to be *annihilated* when the occupation number decreases by one. This description applies not only to the matter field but also to the electromagnetic field: we think of the creation and the annihilation of photons, that is, of energy quanta of the electromagnetic field, when the occupation number of a certain advancing wave increases and decreases by one, respectively.

We arrive now at the final picture of the dynamics of the interaction of the matter field and the electromagnetic field within the realm of quantum field theory, in which energy quanta are created and annihilated. The creation of quanta in one field causes the annihilation of quanta in the other field. Like a seesaw, one goes up when the other goes down. Since there are essentially an infinite number of wave motions of the matter field as well as of the electromagnetic field, creation and annihilation dynamics assume various aspects. A typical example is the dynamics at thermal equilibrium well approximated by quantum statistical mechanics. There, the energy quanta of the various wave motions of the matter field and the electromagnetic field are subject to creations and annihilations restricted only by the conservation law of energy. In other words, a group of energy quanta can be created if their energy compensates for a deficit caused by the annihilation of another group of energy quanta. (Remember the see-saw.) As long as the energy gains and losses balance each other, that is, as long as the conservation law of energy is satisfied, any fundamental process of creation and annihilation of energy quanta may take place with equal probability. Therefore, the creation and annihilation dynamics of energy quanta of the matter field and of the electromagnetic field become totally disconnected from each other, aside from the conservation of energy, and they fall under a statistical description in terms of chance. The creation and annihilation dynamics of the matter field and that of the electromagnetic field take place independently of the details of creation and annihilation processes, provided that the total amount of energy remains constant.

The dynamics of thermal equilibrium is nothing but the random creation and annihilation dynamics of energy quanta subject to the conservation law of energy and the statistical law of equal probability. The random dynamics in thermal equilibrium is said to be disordered. Taking into account only the conservation law of energy and the statistical law of equal probability, we can reduce the framework of quantum field theory to one that is simpler and easier to handle: quantum statistical mechanics.

Umezawa pointed out that it is hopeless to investigate the nature of the dynamics of living matter by means of a framework such as quantum statistical mechanics. Contrary to the disordered creation and annihilation dynamics of energy quanta in thermal equilibrium, the dynamics of the fundamental constituents of living matter manifests order and strong mutual correlation. This might be another aspect of the creation and annihilation dynamics of the matter and electromagnetic fields, which certainly lies far beyond the scope of quantum statistical mechanics. Let us now investigate an aspect opposite that of thermal equilibrium.

Order is literally the opposite of disorder. The term "disordered dynamics" in quantum theory refers to disconnected creation and annihilation dynamics. Hence, it is natural to call certain "connected" creation and annihilation dynamics "ordered dynamics." In what way can such dynamics be "connected"? When does the connection occur? In our daily lives, we usually regard something as "connected" if, and only if, its parts display mutually meaningful behaviors, and we can discern harmony or order therein. For example, imagine the performance of a synchronized swimming team. Each swimmer on the team participates in a harmonious sequence of swimming movements, and the audience perceives therein the existence of a transcendent order or harmony. The audience is impressed not by the details of each swimmer's individual movements but instead by the beauty of the overall harmony and order in the performance. The more the movements of the swimmers are coordinated in a meaningful pattern, the more complete the order appears in the performance.

Suppose the team is training in a sufficiently large swimming pool, and each swimmer performs her own part while maintaining a distance from the other teammates. Then consider these two extreme situations: in the first situation, each swimmer goes through the routine disconnected from the teammates; in the second, the team members move in a synchronized fashion. There is essentially no difference between the two situations in terms of the amount of energy used during the training; however, this is not the case if the swimming pool is so small that the average distance between the swimmers is not sufficient to allow them to perform their parts separately. In the first situation, the team loses energy because each swimmer crashes into teammates and no longer maintains her swimming movement. In the second situation, on the other hand, the team retains its energy because each swimmer rarely crashes into fellow team members and maintains a more orderly performance. The first situation seems to correspond to the disordered dynamics of the fundamental constituents of matter, observed in thermal equilibrium, while the second corresponds to ordered dynamics, a completely new aspect of the nature of matter.

In the case of the actual creation and annihilation dynamics of the energy quanta of matter, the second situation of ordered dynamics is called collective dynamics or a *collective mode*. In the collective mode, standing waves corresponding to a shared elementary energy eigenstate are all equally occupied by energy quanta, and they are so strongly correlated or connected to each other in the collective mode that they manifest a complete harmony, that is, coherence. This concept, often called *quantum coherence*, might provide a key for opening an investigation into the central issues of the two remaining areas of basic research, life and consciousness, within the realm of quantum field theory.

In quantum field theory, matter is thought to be composed of a large number of energy quanta of the matter field. If the creation and annihilation dynamics manifests order, the matter field manifests collective dynamics, and matter is thought to be in an ordered state. In the natural world, it is almost invariably the case that not all matter is in an ordered state but that only some part of matter is. The linear dimension of that part of matter in which the matter field manifests collective dynamics or the collective mode is said to be a coherence length. Depending on the energy eigenvalue of a standing wave in the collective mode, the coherence length may assume macroscopic order. In the collective mode, standing waves corresponding to a shared elementary energy eigenstate are all equally occupied by energy quanta, and a quantum field theoretical investigation reveals this correlation: *the lower the energy eigenvalue of the common elementary energy eigenstate, the longer the coherence length.*

This fact may be well understood if we again imagine a performance of a synchronized swimming team. Recall that the more the movements of the swimmers are coordinated in a meaningful pattern, the more complete the order appears in the performance. Suppose that the movements of the swimmers are intensified, so each swimmer's movement becomes more energetic. Then, it is plausible that the order appears less complete in the performance. That is, it becomes more difficult for each swimmer to follow all the other swimmers' movements, and one might follow only the few team members who are nearby. On the other hand, if the movements of the swimmers are gentle and the energy required for such movements diminishes, it might be much easier for each swimmer to follow all the swimmers' movements. As a result, there appears to be complete order in the performance.

If the collective mode happens to be realized in a certain macroscopic part of matter as a result of the creation and annihilation dynamics of energy quanta of the matter field, a macroscopic part of matter is thought to have the same wavefunction as the shared standing wave. This common wavefunction may be thought of as a wavefunction appearing on a macroscopic scale comparable to the macroscopic region of the collective mode and, thus, it is called a *macroscopic wavefunction.* Here, the macroscopic scale corresponds to the ordinary scale of microns and inches, so the macroscopic objects can be seen with the naked eye or at least with an optical microscope. In other words, the collective mode of the matter field may be described by the new notion of a macroscopic wavefunction. The macroscopic wavefunction is thought to be defined on the macroscopic scale of space and time, that is, on the scale of the coherence length. Therefore, the fact that matter can be well described by the macroscopic wavefunction indicates the possibility that the collective mode of the matter field is realized as the completely ordered creation and annihilation dynamics of energy quanta in the macroscopic spatial region, for which the macroscopic wavefunction is defined. In this sense, the macroscopic wavefunction is also said to be an *order parameter* of the matter field.

As shown in the third part of this book, the collective mode might be the key concept in understanding the true nature of living matter, though it is not restricted solely to living matter. Today, we can identify a variety of collective modes manifest even in non-living matter, such as superconducting material and laser devices. The former is a reflection of a phenomenon typical of the quantum coherence of the collective mode called **Bose-Einstein condensation**. As the occupation number of Bose-Einstein quanta is unlimited, the collective mode of bosons in the matter field could include an unlimited number of bosons, in striking contrast to the collective mode of fermions. The collective mode of bosons, therefore, is much more easily realized than that of fermions, but both of them play important roles in ordering the creation and annihilation dynamics of energy quanta in living matter.

Part Two

Theories of Brain Functioning

Chapter 4

Classical Theory of the Brain

Section 18
The Cosmos of Persona

How can we understand what our brains do? The beauty of rich and sophisticated mental activity arising from an array of billions and billions of simple-looking structures may be aptly compared to the complexity of our universe. The brain is a cryptic cosmos that can be scooped up by our hands, an ever-expanding universe enfolded within an entity of tiny spatiotemporal dimensions. The brain, the source of human intelligence, might be called a cosmos of sapience.

Nothing is more remarkable than **memory** and **consciousness** in the human brain. A long-standing and deep-seated anxiety is common among human beings in the face of these mysterious forces. We wonder: "What is consciousness? What is **mind**? How does memory work?"

Let us start with a brief history of Western thought in response to these questions. Since we cannot open the Akasha chronicle, which is believed to be a supernatural library that only a sage can enter to unveil every truth in the past, present and future, let us trace the written record, beginning with Homelos, a blind Greek poet who lived in the ninth century B.C. He described a vaporous essence of life called *thymos* filling up the whole universe and providing human beings with emotion and intelligence. (*Thymos* is usually translated as "spirit" in English; however, that word is laden with so many meanings that the writers of this text prefer to retain the original Greek term *thymos*.) It may be important to note that Homelos regarded a human being not as a mere union of a material body and thymos, but as a somewhat complex union of body, thymos and *psyche*. *Psyche*, the etymological root of psychology, was perceived by Homelos to be a shabby, unintelligent human soul that does not participate in consciousness but simply activates the body through instinct. Modern psychology, maintaining the belief that the intelligence of a human being can be probed by observing a person's behavior, appears to still be bound by Homelos' fetters and to be looking for the birth of new human science, *thymosology*.

Popular beliefs about thymos persisted from the time of Homelos until the seventeenth century. For twenty-five hundred years, people believed that thymos entered a person's body and endowed him or her with human consciousness. Aristotle of Khalkidike, for example, looked upon the heart as the seat of consciousness. He saw thymos flowing there and creating consciousness. Alkman of Laconia and Hippocrates of Cos believed that the thymos provided the cerebrum with intelligence and gave birth to thought and sense perception. A medical book by Hippocrates, entitled Ancient Medicine and completed approximately in the decade of 420 B.C. advocated the belief that a specific organ called the brain was tightly connected to consciousness. Finally a Roman doctor Galen promoted a doctrine that consciousness arose from the thymos stored within the ventricle in the center of the brain. Although the notion that thymos provided human beings with consciousness was accepted as common sense among European people for twenty-five hundred years after the time of Hippocrates, two Greek philosophers had doubts about that **mind-body dualism**. One was Democritus of Abdera, and the other was Socrates of Athena. Democritus attempted to explain everything—from physical to mental processes—from a mechanistic point of view involving the movement, fusion, and fission of atoms, which were then considered the fundamental constituents of all that exists. He considered consciousness to be nothing but a continuous flow of atoms, which could pass smoothly through anything and move other atoms as they slid past. Socrates disagreed with Homelos and proposed instead that the cerebrum created sense, which generated perception and memory, which, in turn, yielded consciousness. This idea of the young Socrates seems to have been taken over by modern brain science, leaving little room for doubt. It is unfortunate that Democritus's idea has been ignored, though it might be more attuned to the actual nature of the brain and consciousness. As we reflect on the fundamental processes of the brain from the new point of view of **quantum physics**, we cannot help but think of Democritus. Long forgotten in modern science, he finds a place now in the present volume.

The mind-body dualism of Homelos dominated Western thinking for centuries, from the Greek age through the era of the Roman Empire, beyond the Middle Ages, and into the Renaissance period. In the seventeenth century, new ways of thinking were initiated by Gottfried Leibniz and **René Descartes**. The basic idea of Leibniz was similar to that of Democritus. In place of the atom described by the ancient philosopher, Leibniz imagined that a fundamental constituent of the whole of creation, called a *monad*, carried mind and consciousness. In his framework, all creation was nothing but the accumulation and integration of monads. Descartes, one of the fathers of modern science who is well known for setting forth a mechanistic cosmology, proposed that every material body—including every life form—was incorporated into the huge machinery of the universe and governed by the **law of causality**, that is, the law of cause and effect. The human body, too, was regarded as an automatic mechanism, except for this one feature: it had a mind that could both determine certain actions and remember its encounters with the physical environment. A free movement of the human body could be executed due to the existence of a mind with an **ego**. This mind was also supposed to contain a consciousness that registered and retained impressions arising from a mechanistic effect of the physical environment acting on the human body.

Homelos' thymos found a place in Descartes' mechanistic view of the human body. Thymos was perceived to be an extremely rarefied fluid flowing through the cerebrum and the nerves. As it circulated through the body, it linked sense organs as well as muscles to the cerebrum and generated many mechanistic movements. Therefore, mind, present in the thymos, seemed to govern bodily mechanisms. Descartes located mind in the pineal gland, an internal secretion gland in the brain. Consequently, human beings were regarded as nothing more than delicate mechanisms, exceptional only with regard to their ego consciousness. As such, we were nicely fitted to a mechanistic cosmology.

Descartes thought of quite an interesting way to answer these basic questions:

- How are the physical effects of phenomena such as visual, auditory, and tactile sensations stored as memories?

- How can consciousness associate one effect with another?

The mechanism that Descartes envisioned to accomplish these purposes will be presented in the next section.

Throughout history, ideas about brain function and the mechanism of consciousness have evolved along the lines of the prevailing paradigms regarding nature and cosmology. The brain first became associated with mind and consciousness at the time of Hippocrates. Aristotle, Hippocrates, Democritus, Socrates, Descartes, and Leibniz all discussed mind. They pursued the mechanism of consciousness and looked for the source of intelligence within the larger framework of the universe. One cannot help but wonder if their grand spirits have vanished, for as modern science has progressed, the brain has become a subject of medical and physiological analysis and investigation merely as an organ of the body. Its connection to the universe as a whole has become obscure. The split between brain science and physical science is partially responsible for the alienation of the universe of the brain from its home in the macroscopic universe.

Section 19
The Cartesian Mechanism of Memory

In Homelos' mind-body dualism, mental activities such as consciousness and memory were supposed to come from the existence of thymos in a human body, which infused the cerebrum with intelligence. Thymos was seen as a vital element with feelings and thoughts and the capacity to retain memories of past events. Thymos was thought to leave our bodies when we lose consciousness or with the last breath when we finally die. After that, all that remains is a zombie subject to psyche without consciousness and memory. In that model, no mechanism of memory was needed because memory was thought to be carried by an immaterial entity, that is, thymos.

Descartes promoted the idea that mind with ego consciousness existed in the pineal gland and that it controlled the delicate mechanism of the human body by means of thymos flowing from it toward each part of the body. In his view of the universe, every object, including life forms, must be governed by mechanical action satisfying the law of causality in the huge mechanism of the universe. Human beings were regarded as nothing more than delicate mechanisms, exceptional only with regard to their ego consciousness. As such, we were neatly fitted to a mechanistic cosmology. Accordingly, the role of ego consciousness, regarded as an exception within the realm of the Cartesian mechanistic cosmology, was minimized. Descartes seemed to want to change the exceptional character of the memory process for that very purpose. He thought of a delicate mechanism of memory by which sensory impressions are imprinted and recalled in the cerebrum. It is very interesting as a model of a memory mechanism invented for the first time by a human being. Moreover, it is noteworthy that this model grasps the process of memory imprinting and recall quite well, although his basic idea is very simple. What kind of delicate mechanism did he conceptualize?

Descartes regarded the nucleus of consciousness as an ego emerging in the pineal gland. He thought that thymos created intelligence by flowing in the cerebrum and that ego perceived the outside world through various organs. Descartes maintained that awareness of the outside world occurs when a sense organ is affected by the body's contact with the external environment; that is, an element in the physical world acts on the body to cause a sensory response. The mechanical action, transformed into thymos in the sense organ, flows through the nerves and up to the cerebrum. Hence, ego consciousness emerging in the pineal gland perceives the outside world by scanning the flow of thymos transmitted from the sense organ to the cerebrum. An impression, i.e., a perception of the outside world, is preserved as a memory and recognized continually thereafter by ego consciousness. Where in the brain does this process of memory imprint and recollection take place?

Descartes invented a rather delicate memory mechanism consistent with his own mechanistic view of the universe. First of all, he thought of brain tissue as having a textile-like structure of woven nerves. It might be helpful here to picture a stiff hemp cloth crumpled into a ball. Thymos transmitted from a sense organ through nerves flows into the brain tissue made of nerve textiles, and, at this point, Descartes made a very important assumption: Brain textiles, subject to the flow of thymos, change in such a way that the texture becomes thinner if thymos flows there and thicker if not. Furthermore, a thin texture remains so for a period of time, even after thymos stops flowing there; the more frequently the texture is thinned, the longer it remains that way. Maintenance of thin textures is memory. So said Descartes.

It is worthwhile to note here that thymos moves more easily from a sense organ into brain textiles if the texture has been thinned previously. Although new inflow of thymos makes textures thinner, it is more likely that thinner textures are already there and maintained as memories of mechanical effects caused by prior contact with the outside world. Hence, thymos newly arriving from a sense organ places a new and different pattern upon a previous pattern of brain textiles that already have thin textures. This recurring stream of thymos through textiles with overlapping patterns extends the duration of those textures. Furthermore, thymos flowing anew from a sense organ creates a new pattern of thinner textures as it interacts and interferes with the previously thinned pattern of brain textiles. The new impression of a sensory experience collates with the preexisting memory, and this compound impression is stored as a new memory.

For Descartes, memory was nothing more than an impression of the outside world woven into the pattern of brain textiles, which are actually nerves. If we take the aforementioned picture of a hemp cloth, we can imagine a beautiful pattern drawn by thinly woven textures. This pattern represents an impression of the outside world, that is, a memory of nature obtained through sense organs. A trace of memory maintained in the cerebral tissue as a qualitative change of a certain physiological factor has been sought by scientists even in the twentieth century, but thus far no one has found the elusive *engram*, the existence of which was first proposed by Descartes in the seventeenth century. In the Cartesian literature, an engram is a ***moiré*** pattern of thin textures woven into brain textiles made of nerves. It is, indeed, a ripple mark in which all impressions of the outside world overlap and intersect each other.

From the Cartesian point of view, a person can be conscious of an impression maintained in the brain as an engram because a mind bearing an ego—that is, a self-conscious intelligence—exists in the pineal gland. This mind was thought to have a life of its own, independent of causal action from an outside physical agent. From it issues thymos. Descartes did not provide a detailed account of the mechanism that supposedly secretes thymos. He merely postulated that this nebulous mind in the pineal gland emits thymos, which scans the pattern of textures woven in brain textiles as it streams through those textures. Quite a skillful mechanism, isn't it? Of course, that account of consciousness might be nothing more than a naive fantasy since the theory was based on a medical model conceived in the seventeenth century. However, the fundamental concept of the Cartesian memory mechanism is still alive in another form within a **quantum theory** of the fundamental processes of memory. We will come back to it again when we discuss that quantum theory of the brain.

Section 20
The Dawn of Modern Brain Physiology

During the seventeenth century, Descartes and Leibniz laid the groundwork for a mathematical and mechanical view of nature, which shaped the thinking of many scientists and philosophers and culminated in the work of **Isaac Newton**. Their epoch-making theories spurred remarkable developments in optics, mechanics, hydraulics, bridge engineering, artillery, and navigation. The Science Academy, founded in Europe, contributed greatly to the technological advances made possible by **classical physics**. Unfortunately, brain research, particularly the investigation of a mechanism of consciousness arising from a brain function, merely roamed on the outskirts of the Cartesian world view until the beginning of the nineteenth century.

At the end of the eighteenth century, Frantz Gall, a doctor in Vienna, set forth a phrenology that associated consciousness with the cerebral cortex. It was his belief that in many cases a distribution map of the skull's development correlates closely with certain characteristics of consciousness, though he could provide no experimental evidence. However, that was the first time the idea was presented that consciousness arises from the cerebral cortex that covers the brain surface and affects most intensely the development of the skull. Although Gall's phrenology did not provide us with a deep understanding of the mechanism of consciousness in the brain, it may be rated highly on the point that people finally took notice of the cerebral cortex. In fact, in the latter half of the nineteenth century, portions of the cerebral cortex were found that are linked to speech and motor functions; thus, they are called the speech and motor areas.

The late nineteenth century was also a remarkable epoch during which investigations into electricity and magnetism resulted in the telegram, the telephone, and so on. The **theory of electromagnetism** completed by **Maxwell** was believed to be the crowning accomplishment of the mechanistic view of the universe that originated with Descartes and led to Newton's system of mechanics in the seventeenth century. Believing that the essence of physics, including thermodynamics and the kinetic theory of gases, had been thoroughly studied, physicists became more attentive to the study of the inner structure of **matter** as the next terra incognita. Modern chemistry bloomed on the basis of Dalton's atomic theory in 1808, and a materialistic view of nature was accepted gradually, stimulated by the success of chemistry in analyzing and synthesizing matter in terms of **atoms**. Within that historical context, modern brain physiology was born. Its labor pains are thought to have started with Wilhelm von Waldeyer and Ramón y Cajal at the beginning of the twentieth century, when they thought of nerve cells as the fundamental constituents of the brain. At that time, the mechanistic view of nature, based on Newtonian mechanics and **Maxwellian electromagnetism**, began to show signs of failure, though until then it had seemed impregnable.

Waldeyer and Cajal proposed that brain tissue consists of numerous nerve cells called *neurons* bordering each other. Prior to their discovery, the brain had been regarded as a continuous tissue made of textiles of nerve fibers, as claimed by Camillo Golgi and others. Waldeyer and Cajal's idea that brain tissue might be an assembly of independent cells, like every other physical tissue, could not be verified by observation through the optical microscopes available in those days. The microscopes needed to verify their theory did not exist until a quantum theory that assigned a **wave** property to an **electron** provided the basis for the electron microscope.

The long labor pains started with Waldeyer and Cajal's neuron doctrine. Then came a man who served as a midwife to deliver modern brain physiology. He was Charles Sherrington of England. He compared the result of a physiological study of the electrical characteristics of nerve fibers with that of a behavioral study of reflex motions of vertebrate animals and concluded theoretically that nerve cells in brain tissue border on neighboring ones at extremely small spaces. The electrical characteristics of nerve fibers ensure that the propagation speed of electrical signals is much higher than the actual transmission speed of stimuli in the reflex motion. Furthermore, the propagation of electrical signals along a nerve fiber was shown to have no directivity—in physics and electrical engineering, something is said to have "directivity" when it flows in one direction but never in the opposite direction—though the transmission of stimuli in the reflex motion has evident centripetal directivity. Those facts led Sherrington to discern the existence of extremely small spaces between the neurons participating in the line of transmission; that is, he noted that neighboring nerve fibers made of connecting nerve cells have extremely small gaps between them. This interval or gap, named the *synapse* by Sherrington, became the heart of subsequent brain science.

That breakthrough occurred in the year 1906, though the existence of the synapse was not widely known until the publication of Sherrington's famous book *The Integrative Action of the Nervous System* in 1947. The membrane electric potential of the nerve cell had already been observed by then. It was also known that active electric potentials called neural impulses play an important role in the fundamental processes of the brain. In 1952 Alan Hodgkin and Andrew Huxley thought out a set of differential equations that can reproduce qualitatively the fluctuations of membrane electric potentials. Earlier Warren McCulloch and Walter Pitts proposed an electric-circuit model of the neuron as a multi-input and single-output threshold logic device subject to the rule of decision by majority. Of course, electron microscopes invented by Ernst Ruska were also very helpful in the direct verification of synapses.

Thus, modern brain physiology may be said to have been opened by the recognition that the brain is an assembly of untold numbers of nerve cells. Memory as well as consciousness has been thought to arise from the mutual workings and interactions among nerve cells. To give credit where credit is due, we should note that by 1911 Cajal had already presented his idea that memory is carried by neurons and their mutual joints. It was Karl Lashley who squarely faced this problem in modern brain physiology: Where can traces of memory, that is, engrams, be found?

Section 21
The Embryology of the Old Brain

Let us examine various aspects of our brain. The complicated structure of the human brain can produce highly developed mental activities such as consciousness and memory. Although it would be very interesting to actually see the fundamental processes of the mind, they lie concealed behind the complex material structure of the brain and are not directly observable. Let us then reflect on various stages of development.

At the earliest stage of development, a person is no more than a single biological cell. The life of a person as a new carrier of consciousness starts with a male sex cell, that is, the spermatozoon, uniting with a female sex cell called the ovum or egg. The spermatozoon penetrates one spot of the ovum membrane. At the moment of this invasion, fluctuations occur in the concentration of calcium ions and spread over the ovum membrane as a fertilization wave, thus changing the ovum membrane so that it serves as a barrier against other spermatozoa. This fertilized egg slowly rolls over the inner wall of the oviduct as cell division occurs. About seven days after conception, the egg lies buried in the uterine epithelium and, at that point, is called a blastula or embryo. Clusters of cells called the entodermal, mesodermal and ectodermal parts develop within the blastula. A portion of the cells of the ectodermal parts become the long and narrow, primitive tissue called the neural plate, and the remaining cells become skin tissue.

This process occurs within two or three weeks of fertilization. Two or three days after that, the two edges of the neural plate develop as the cells in that region pile up, and finally they come together at the meridian to form a tubular tissue. This is called the neural tube, and it serves as the root of the central nervous system. (Of course, the neural tube is a primitive tissue found only during the developmental period of a vertebrate animal; no such tissue can be found in invertebrate animals.) A group of cells located on the outer edge of the neural tube forms the neural corona. These cells settle between the central nervous system arising from the neural tube and the skin tissue arising from the remainder of the neural plate; they become the root of the peripheral nervous system. The peripheral nervous system thus emerges from the neural corona.

Once the neural tube is formed, the epithelium cells in its inner wall reproduce and form two kinds of cells: One is a nerve cell called a *neuron*; the other is a *neuroglia cell* called simply a *glia cell* or *glia*. The primitive epithelium cells in the neural corona also proliferate through cell division and become neurons as well as non-nerve cells called Schwann cells. The latter will form the nerve sheath.

Let us first look at the neurons and glia produced in the neural tube. Cell reproduction occurs most prolifically in a specific area of the inner wall at one end of the neural tube: it is this part that will form brain tissue subject to sophisticated metamorphoses. At the same time, the eyes develop in the neighboring skin tissue. The remainder of the neural tube, not engaged in cell reproduction, maintains its tubular shape while becoming the vertebral column. This, of course, is the case for vertebrate animals. In invertebrate animals, a cluster of neurons arising from the embryo turns into a brain. Among the vertebrate animals, the metamorphoses in the front portion of the neural tube are most remarkable in humans; this accounts for the highly differentiated character of the human brain. Neurons and glia produced in the front and inner wall of the neural tube move like amoebae to the outside and build brain tissue. At that time, glia cells play the role of trailblazers and cut paths through which neurons follow. This migration of neurons and glia is the initial step in the process of differentiating the three parts of the brain: forebrain, midbrain and hindbrain.

Meanwhile the neurons and Schwann cells produced in the neural corona also move like amoebae toward the outside, where they line up in rows alongside various body tissues that are developing simultaneously and form the peripheral nervous system. Later they will link the central nervous system to various organs in the body. Their root, the neural corona, continues to grow and turns into neural ganglia, which serve as the contact points between the peripheral nervous system and the central nervous system. Neurons and glia multiplying at the front end of the neural tube perform a variety of tasks. For instance, they form two partitions known as the telencephalon and the diencephalon in the forebrain, they merge the diencephalon with the midbrain, and they differentiate the cerebellum from the hindbrain. At that time, some of these migrating cells from the neural tube advance to each neural ganglion and form connections to the peripheral nervous system. The neurons and glia produced in the remainder of the neural tube (which develops into the vertebral column) also move like amoebae to specific neural ganglia and help to complete the connections to the peripheral nervous system.

At this stage of development, the brain and the nervous system of a human appear much the same as those of lower animals. The peculiar complexity of the human brain seems to emerge afterward. Such highly advanced development is realized in both the telencephalon and the diencephalon about two months after fertilization.

Section 22
Birth of the New Brain

The human brain develops to the stage of the so-called primitive, reptilian brain within about two months after conception, and it continues to develop spectacularly not only during the remaining seven months in the womb but also after birth. This development is especially remarkable in both the telencephalon and the diencephalon. In the latter, fine systematized regions called the thalamus and the hypothalamus are eventually formed. The thalamus and the hypothalamus mediate between the telencephalon and the other nervous systems. These regions of the diencephalon control the internal organs through autonomic nervous as well as internal secretion systems. Of course, part of the hindbrain belonging to the old brain also plays the role of mediator between the telencephalon and the peripheral nervous system, and it controls respiration as well as heartbeat. The cerebellum, formed by means of differentiation, is a mediator, too; it controls physical movement by connecting the peripheral nervous system to the telencephalon.

The telencephalon seems somewhat distinguished from the other divisions in the brain. Indeed, it comes to control the highly advanced mental activity and sophisticated forms of intelligence of the human being. Let us look at its developmental process. Following the two-month formative period after fertilization, the forebrain begins to develop to the right and to the left of the enlarged front end of the neural tube. Glia produced in the inner wall of the neural tube swell and radiate outward to the right and to the left, leaving a cave in the brain, which is called the ventricular space. Neurons produced in the inner wall also move along this radial pattern of glia, then, after arriving at the outermost area, leave the glia and lie in a pile atop previously settled neurons. This layered tissue on both sides of the forebrain is the telencephalon. It grows up rapidly and covers both the diencephalon and the cerebellum as the neuronal tissue wraps the right and left sides of the ventricular space, like lava spewing both directions from a crater, piling atop previously settled lava, leaving a cave in the mountain, and enlarging the volcano. Hence, the telencephalon comes to be called the cerebral cortex. In this cerebral cortex, the part formed earlier is called the allocortex, and the newly formed part that overlays the allocortex is called the neocortex. The neocortex comprises approximately ninety percent of the human cerebral cortex. The hippocampus, which is a specific area in the allocortex, is known to control memory, though memory itself is thought to be stored in the neocortex. That is, the hippocampus manages the logistics (i.e., the retrieval and the recovery) of memory, which is stored in the neocortex. These two sections of the cerebral cortex are parted to form the right and left hemispheres or the right brain and the left brain.

The human cerebral cortex thus complete performs the highly advanced functions of memory and consciousness unmatched in any other species. As we mentioned earlier, Gall's phrenology led researchers to further investigate the relevance of brain differentiation to specific mental functions. Experiments with electric current in the second half of the nineteenth century also indicated that mental functions might be localized in different parts of the brain. However, it was not until the first half of twentieth century that modern cerebral physiology provided convincing evidence that the cerebral cortex generates memory and consciousness.

Section 23
A Star Map of the Brain

The investigation of our cryptic cosmos, the human brain, has finally joined modern science in the twentieth century. When we compare concepts of the brain and the mechanism of consciousness that were widely held from ancient times until the late nineteenth century with our current scientific understanding of brain functions, we see a disparity similar to the one between astrology and astronomy. In order for modern astronomy to cast off time-honored astrology, a star map was needed. Tycho Brahe and John Flamsteed provided that in the year 1725. The same kind of localization was needed for the study of the brain, especially for the cerebral cortex. Indeed, regions strongly linked to speech function were discovered in the left hemisphere of the cerebral cortex by Pierre Broca and Karl Wernicke during the second half of the nineteenth century. They induced movement in each part of the human body by administering an electric impulse to the cerebral cortex during surgical operations. Those experiments revealed that various regions in the cerebral cortex are linked to motor functions.

Just as Tycho Brahe and John Flamsteed provided a star map, Karl Lashley in the United States and Walter Penfield in Canada provided a brain map. Penfield, a brain surgeon, established a distribution map of the functional locations in the cerebral cortex, which may be understood as a star map of the brain cosmos. He accomplished this by observing a patient's reaction to a weak electrical impulse given by a microelectrode attached to different parts of the cerebral cortex during a craniotomy. (A local anesthesia killed the pain and still enabled the patient to tell Penfield what he or she saw, heard, or felt when a weak electrical impulse was given.) The result was that each part of the patient's body moved unintentionally when a corresponding part of the cerebral cortex was chosen as a spot to attach a microelectrode and a weak electrical impulse was given. Furthermore, the patient could recall peculiar scenes, voices, or songs when weak electrical impulses were administered through microelectrodes to various parts of the cerebral cortex.

Let us distinguish the four parts of the cerebral cortex according to their relative locations: they are the front (frontal), upper (parietal), back (occipital), and side (temporal) parts or lobes. Although the classification looks obscure at a glance, the lobes are separated by deep grooves and formed by differentiation when the telencephalon develops into the cerebral cortex.

We shall start with a brief look at the frontal lobe. A belt-like region spreading from right to left at the rear of the frontal lobe is named the motor area or the motor cortex. As the name implies, this is the area that activates the motion in each part of the human body. Here, aligned from right to left, is an array of dozens of small regions, each of which is assigned to a fine motion in a different part of the body. For example, the motor activities of the five fingers of the right hand are assigned to five adjacent regions in the left part of the motor area. Similarly, those of the five toes of the right foot are assigned to five other adjacent regions in the left part of the motor area. Those small, aligned regions of the motor area may be compared to the keyboard of a piano. Each key is assigned to a different musical interval. It may be of certain interest to note that each region assigned to each finger is much larger than that assigned to each toe. This fact may explain why we can control the motion of the hand much more precisely than that of the foot. In front of the motor area we find another region, called the premotor area, which coordinates the motor activities of a human body and produces the systematized bodily motion. Coming back to the keyboard analogy, the piano cannot produce good music when many keys are hit at random. We need a musical score to coordinate the way we hit the keys so that the resulting sequence of musical notes will be appealing music. Similarly, systematized bodily motion requires that various motor activities of a human body be coordinated by something analogous to the musical score. It is the premotor area that plays the role of a musical score to coordinate the motor activities of a human body and to produce systematized bodily motion.

The front line of the frontal lobe is called the prefrontal area or the frontal association area, which is thought to produce sophisticated mental activity such as emotion, will and creativity. Mainly restricted to the left hemisphere, a specific region at the side of the frontal lobe controls speech function, as discovered by Pierre Broca in the middle of the nineteenth century. Today that is called the prespeech area or Broca speech area.

Next we shall look at the parietal lobe: the front line of it, which borders the motor area in the frontal lobe, also forms a belt-like region spreading from right to left; this region is called the somatosensory area. There, again aligned from right to left, is another array of dozens of small regions precisely assigned to the sense of touch in each part of the body, just as the motor area controls various movements.

Both the occipital and the temporal lobes manifest activity to control the senses, but vision (sight), the most complex sense, is strongly linked to a region in the occipital lobe called the visual area. The auditory sense is allocated to the auditory area located in the temporal lobe. Karl Wernicke in the second half of nineteenth century found another region that is strongly linked to speech function; this auditory area is also localized in the temporal lobe in the left hemisphere. This temporal lobe is also believed to have some connection with highly advanced functions such as memory and decision-making.

The localization of different functions distributed over the cerebral cortex, as established by Penfield, became a model for investigating the workings of the cerebral cortex; in fact, it became the foundation upon which modern brain science was built from the beginning to the middle of the twentieth century. But does the cerebral cortex really localize its functions and specify its working in a highly differentiated manner, as Penfield claimed? Remember that he reached this conclusion of functional localization by giving weak electrical impulses via a microelectrode to the cerebral cortex during a craniotomy. Was his conclusion premature? Were the cooperative actions of other parts of the cerebral cortex perhaps overlooked? Penfield's doctrine of functional localization in the cerebral cortex should be reconsidered from a different perspective, one that takes these questions into account. A new step in this direction has been taken quite recently as a result of the development of more sophisticated technology, specifically Positron-CT (computer tomography) or PET (**positron** emission tomography). In the near future, a completely new type of functional distribution might be discovered in the cerebral cortex and, we expect, the star map of the brain will be redrawn.

In the meantime, the debate over functional localization continues. Penfield based the distribution of functional localization in the cerebral cortex on the results of his microelectrode experiments performed during brain surgery. During an operation performed to alleviate epilepsy, he discovered a region in the temporal lobe that appeared to be strongly linked to memory function. Based on that finding, he claimed that the memory function is also localized in the cerebral cortex. That was in the year 1958. At the same time, the American neuropsychologist Karl Lashley was also pursuing the trace of memory (i.e., engram) in the cerebral cortex. For more than thirty years, he devoted his energies to experiments with rats and monkeys and finally concluded that memory is not localized in a specific region of the cerebral cortex but is distributed over the whole cerebral cortex. While conducting experiments such as maze-learning in rats and visual pattern-recognition in monkeys, he divided each small piece of the cerebral cortex that stored memory after learning had taken place to determine if the memory contents were also broken. Here are the results: he found no indication of the alleged localization of either memory or brain functions; damaging specific cortical pieces did not annihilate the memories or the functions supposedly associated with them. Lashley did observe, however, that the greater the damage to the cerebral cortex, the greater the disorder. Thus, during the middle of the twentieth century, brain science split into two camps, each maintaining its own doctrine: that of functional localization (Penfield) or that of nonlocalization of the cerebral cortex (Lashley). It was the confrontation between the doctrines of locality and nonlocality that triggered the explosive progress of modern brain science.

Which of those seemingly opposite experimental results will be declared the winner? Penfield's or Lashley's? Can both of them be right, possibly providing us with different aspects of brain functioning from different points of view?

Section 24
Brain Waves

At the same time that Lashley and Penfield were working in the United States and Canada, respectively, to resolve the debate over nonlocalization versus localization of functions in the cerebral cortex, Hans Berger in Germany discovered weak electric potential oscillations recorded by an electrode attached to the human scalp. Berger verified that the electric potential fluctuates characteristically in accordance with variations of consciousness. That was in the year 1929. Since a wavy pattern was obtained by monitoring scalp electric potential oscillations over time, it was called a brain wave. (The more professional or technical term is electroencephalogram, abbreviated to EEG.) It is generally difficult to measure the electrical activity of the cerebral cortex directly by placing microelectrodes on it during a craniotomy, whereas brain waves can be observed fairly easily by attaching microelectrodes to the scalp. Thus, since Berger's discovery, brain waves have been studied extensively as unique scientific indicators of both brain activity and states of consciousness. The relationship between brain waves and states of consciousness has been of particular interest to researchers. Extensive experimentation has shown that the number of oscillations in a unit of time, that is, the **frequency** of brain waves, is more clearly indicative of brain activity and of states of consciousness than is the **amplitude** of brain waves.

There are the four standard classifications of wave frequencies, namely, *beta*, *alpha*, *theta*, and *delta*. Those with a frequency higher than thirteen hertz, that is, thirteen or more oscillations per second, are classified as beta waves. Those ranging between eight and thirteen hertz are called alpha waves, from four to seven indicate theta waves, and from one to three delta waves. In general, alpha and beta waves are recorded during the waking state: the latter become dominant if the subject has his or her eyes open and is conscious of external stimuli; the former emerge if the subject becomes calm, with his or her eyes closed. Theta waves are observed during the period of shallow sleep when one falls asleep, while delta waves are recorded during deep sleep. Of course, these are merely rough classifications of broad states of consciousness indexed phenomenologically by the frequency of brain waves, and many exceptions have been found. The most notable exception is that during meditation in an awake state, Zen priests have registered alpha and theta waves.

By observing brain waves, it is possible to study changing states of consciousness somewhat objectively, though the results are only rough approximations within four broad categories. It seems worthwhile to note here that the scalp electric potential oscillations called brain waves are assigned to brain activity and states of consciousness only phenomenologically. In other words, *we have no definitive proof that brain waves represent the workings of the brain or the functioning of consciousness, though there is a correlation between those phenomena.* To be sure, these questions remain unanswered:

- What is a brain wave?
- How does it arise?
- How do brain waves relate to brain activity and to states of consciousness?

It is widely believed that brain waves are a record of the electric potential distribution on the scalp, arising from the summation of the electrical activity of all the nerve cells over a vast region of the cerebral cortex. Starting from this premise, one might easily surmise that the fluctuating electrical activity of each nerve cell contributes to the summation over a vast region of the cerebral cortex, and the overall effect is that they cancel each other out. That is, a summation of the electrical activity of all the nerve cells results in no electric potential distribution on the scalp as a whole if the fluctuating electrical activity of all the nerve cells remain asynchronous. Asynchronous membrane electric-potential activity does not produce any measurable electric potential distribution on the scalp. Therefore, any measurable change in the electric potential recorded by scalp electrodes, that is, the existence of any brain wave, requires the summation of *synchronous* electric potential activity among nerve cells over an area of at least six square centimeters in the cerebral cortex.

Today, of course, many scientists assume that brain waves result from an action initiated in the cerebral cortex by a yet unknown mechanism. Many researchers are working from this assumption to obtain a clearer view of brain functioning. With the help of sophisticated technology, they are able to map out regions in the cerebral cortex where brain waves are generated. They use a frequency filter to isolate one class of brain waves, say alpha waves, and a computer to calculate and display the intensity distribution of alpha waves in the cranium. A few scalp electrodes can simultaneously measure a restricted number of data to be processed by a computer. The intensity distributions of beta and theta waves can be mapped out simultaneously by means of multiple frequency filters, and certain aspects of brain activity, such as the predominance of right or left side engagement, can be seen in the cerebral cortex. This research offers more detailed evidence of the correlation, revealing, for example, that one type of mental activity, elementary mathematics, is correlated with the production of brain waves on the left side of the cerebral cortex.

The cerebral cortex is divided into right and left hemispheres from the early stage of fetal development. This question naturally arises: Do the right brain and the left brain perform the same functions? Or do they play different roles? The first person to offer a well-supported answer to this question was the psychobiologist Roger Sperry in the United States. He performed various experiments on epileptic subjects whose callosal body (corpus callosum) joining the right and the left hemispheres had been severed, and he verified that each hemisphere takes partial charge of brain function. For example, the left hemisphere appeared to preside over rational, intellectual activities, such as language, logic and algebra, while the right hemisphere seemed to govern mental activities involving imagery and sensitivity, such as visual arts, geometry, music, and intuition. That was in the 1970s. Thereafter, the investigation of the distribution of functions between the right and left hemispheres of the cerebral cortex, namely, the right and left brains, received a tremendous amount of attention from researchers and the medical community in general. Today, no doubt, the existence of a rough division of brain functions will be clarified by mapping the intensity distribution of brain waves, as mentioned before. There are, of course, exceptions in which the usual distribution of functions is not apparent. For example, one of the winners of a university mathematics contest in the United States was once given a diagnosis of hydrocephalus, and the person's left brain was found to be missing. The winner got the highest score in solving challenging mathematical problems at the university level relying only on the right brain! Nevertheless, the complementary roles of the right and left brains might be discovered in the near future, provided that the meaning of the existence of brain waves can be clarified through further investigation with more sophisticated measurement and data processing technology than that which is currently available.

Section 25
The Classical Theory of the Brain

Although Sperry found evidence that the right and left hemispheres of the cerebral cortex each partially control highly advanced mental activities, it is not yet an established fact that either the right or the left brain dominates particular mental processes. Therefore, the debate continues over Penfield's locality doctrine of cerebral functioning versus Lashley's nonlocality doctrine. Perhaps this conflict cannot be resolved by research solely within the realm of cerebral physiology.

With physiology as the foundation of modern medicine, there is a tendency to explore the origin of physiological phenomena in terms of the accumulation of fundamental physical and chemical processes on the microscopic scale. Recent remarkable research in the area of **molecular biology** has certainly contributed to this trend and has given rise to a school of cerebral physiology with a similar methodology. In this vein, researchers first study the fundamental physical and chemical processes of the basic constituents of the brain, that is, the nerve cells, and then attempt to understand the mechanisms of consciousness by compiling the data. In other words, they are trying to piece consciousness and memory together by assembling the physiological fragments of brain tissue. That is rather like trying to create a whole picture of an elephant by piecing together fragmented images of the animal obtained by a number of sightless persons touching different parts of the elephant. From this standpoint, the great pioneers in modern brain research—Penfield, Lashley, and Sperry, for example—seem to be following the classical tradition of brain physiology.

Chapter 5

Physiological Theory of the Brain

Section 26
The Brain as Matter

It might be true that **consciousness** arises because of the existence of **memory**. A human being usually responds to new situations on the basis of cumulative experience stored in memory. The impact of each "collision with the world" is stored as a fresh memory and used as a reference when one is deciding how to respond in a new situation. Might the higher order of mental activity called consciousness in humans be nothing more than brain work that integrates memory with new impressions gleaned from external stimuli? Put another way, is consciousness simply a continual process of collating the stored records of one's past reactions with a succeeding chain of events for the purpose of determining an appropriate course of action? As you can easily see if you close your eyes, the richness and diversity of memory is astonishing. By what mechanism is such a cryptic force as memory realized in our brain?

Classical theories about the brain, starting from ancient myths and continuing into modern cerebral physiology, have not uncovered the mechanisms of memory and consciousness as they relate to brain functioning. Perhaps the search for the elusive source of these mental phenomena is far beyond the scope of empiricism or phenomenology. We are certainly approaching a turning point in our analyses of the brain as we delve deeper into its material constituents. Our understanding of the behavior of **matter** in the sub-microscopic world has increased significantly as a result of the discovery of **quantum mechanics** during the first half of this century. Subsequently, various precision measurement techniques—for example, electron microscopy, NMR-CT (nuclear magnetic resonance computer tomography) and PET (**positron** emission tomography)—have been created. Concurrent with these technological advances, **molecular biology** and quantum biology have come into being; in these sciences, life phenomena are studied from a purely materialistic point of view, that is, in terms of masses of **atoms** and **molecules**. In order to uncover the true nature of the universe as it evolves over time, scientists have needed to understand the fundamental behavior of matter in that process of change. The same can be said about our inquiry into the cryptic cosmos of the brain.

Section 27
The Cell

As explained in Section 21 of Chapter 4, two kinds of cells, neurons and glia, emerge as the epithelium cells reproduce in the inner wall of the neural tube. Shifting their positions outward, these cells form the brain tissue. As the blood vessel system develops, glia encircle the cells forming capillary vessels in the head and the cervix. They selectively absorb only glucose, amino acids, and oxygen through the endothelium cell membranes of the capillary vessels. There are no gaps between the endothelium cells of the capillary vessels in the brain, and only the molecules transmitted from the endothelium cells to glia cells can pass through cell membranes. Therefore, macromolecular substances that cannot penetrate the endothelium cell membrane of the capillary vessels do not become part of the brain tissue. A system called the blood-brain barrier plays the role of a gatekeeper, protecting the brain against an invasion of noxious substances and preventing neurotransmitters, which must pass through the brain, from flowing out into the blood. This protective mechanism isolates the brain tissue from the external environment and, thus, maintains its stable functioning.

Neurons are embedded in groups of neuroglia cells that surround the system of capillary vessels. Neurons are thus screened from the system of capillary vessels by the neuroglia cells, and so receive glucose, amino acids, and oxygen only through the membranes of neighboring neuroglia cells. Within the cells of neurons and glia, chemical energy produced by oxidization of glucose is stored in the form of a molecule called ATP (adenosine triphosphate). Those ATP molecules "fuel" the neurons and glia, so to speak; that is, they provide the neurons and glia with the **energy** needed to perform their physicochemical activities. Various physicochemical processes stop if, for one reason or another, glucose and oxygen are no longer supplied. There is clinical proof that one loses consciousness within ten seconds in such a circumstance; therefore, it is conceivable that human consciousness functions on the basis of physicochemical reactions both inside and between neurons and glia in the brain.

An energy supply in the form of glucose and oxygen is indispensable to the maintenance of brain functioning, though alone it is insufficient to keep the brain stable. Most of the organic macromolecules constituting nerve cells and neuroglia cells, which form the framework of the brain, disintegrate within two weeks and are replaced by new ones. The constituent molecules of those cells cannot be supplied from outside the brain tissue because, you will recall, the blood-brain barrier prevents them from entering; therefore, they must be produced in the brain tissue from amino acids that can pass through the gateway. The neuroglia cells play a role here. Inside the neuroglia cells that surround the capillary vessels, amino acids facilitate protein synthesis, which produces the molecular materials required to reconstruct the cytoskeletons of neurons and glia. Here we might be seeing a general characteristic of life systems or an essential quality of life, namely that every living body functions as a unified whole in the long-term, even though the elements of its overall structure are relatively short-lived and are continually being replaced by new ones. This behavior, which is chemically based, is called metabolism. The brain metabolism is thought to be due mainly to the workings of the neuroglia cells.

Thus, it becomes clear to us that brain tissue is subject to metabolism, just like other bodily tissues, and its constitutional molecules are continually being replaced by new ones. Memory, however, is assumed to arise from brain functioning that remains stable over a long period of time. Taking this continuity of memory into account, we might not find the trace of memory, i.e., the engram, in the constituent molecules of the brain seen as a huge cell assembly of neurons and glia, but rather in the brain's macroscopic structure, which remains invariant through metabolism. If so, what kind of macroscopic structure of the huge cell assembly exists in the brain? Before proceeding to this issue, let us take a look at the cellular structure of the brain in terms of its neurons and glia.

Section 28
The Brain Cell and Cell Membrane

A brain tissue cell, such as a neuron or a neuroglia cell, manifests a spatial structure like other cells in which cytoplasm is confined within a limited region by a membrane. Cytoplasm is a complicated mixture of about seventy percent water, about twenty percent protein, and the remainder many other types of molecules. Cytoplasm contains a variety of microorgans that are functionally differentiated and characterized by specific protein molecules. The largest among the microorgans is the cell nucleus, which isolates the organic macromolecule storing heredity information, that is, DNA (deoxyribonucleic acid), by enclosing it in an internal membrane called the nuclear membrane. There is only one nucleus in a nerve cell or glia, whereas the second largest microorgans, called mitochondria, are numerous. These mitochondria produce the molecules of adenosine triphosphate (ATP) in the cell by burning (i.e., oxidizing) glucose. In other words, mitochondria generate the energy needed for the activity of brain cells such as neurons and glia. The microorgans producing various frame molecules are endoplasmic reticula and ribosomes. The reticula synthesize phospholipid molecules that form the structure of the cell membrane, and the ribosomes make protein molecules from amino acids.

Serving all these purposes, cytoplasm is indispensable to both the steady maintenance and the activity of the brain cell, though higher order brain activity, such as memory and consciousness, certainly cannot be extracted from a single brain cell. Brain tissue performs such functions only if an enormous number of brain cells, such as neurons and glia, are assembled. Thus, the functioning of the brain as a cell assembly of neurons and glia can be thought to arise, not from the specific events taking place in the cytoplasm of each cell, but from the cooperative work among the cells. Taking this fact into account, we see that even though cytoplasm plays the fundamental role of maintaining the brain cell itself, it seems to be less important in realizing brain functions since the cytoplasm in a single cell is isolated from that in other cells by cell membranes. The cell membrane is more fundamental. A detailed picture of cell membranes follows.

In the early 1970s, the *New England Journal of Medicine* published an interesting series of articles on the subject of biological cells. In the final article, Lewis Thomas, an American pediatrician, likened the Earth to a biological cell (Thomas 1975). Unrelated to Jim Lovelock's *Gaia* hypothesis pertaining to the global life sphere, Thomas's analogy is more concerned with the intrinsic qualities of cell membranes. That is, when the Earth is compared to a biological cell, the part corresponding to the cell membrane is the atmosphere; that is the world's largest membrane. The atmosphere never appears uniform. Dispersed in various areas of the atmosphere is the water vapor that forms clouds, which release the rain that flows over the surface of the Earth and furnish the cytoplasm, that is, the sea and land, with electrical energy in the form of occasional thunderbolts. The atmosphere, blocking the rays of the sun and **cosmic rays**, keeps the environment of the Earth stable by protecting it against harsh attacks by high-energy radiation. Any kind of invasion from extraterrestrial space must pass through the Earth's membrane, that is, its atmosphere. Of course, we have to go through the atmosphere, too, when we visit other planets, such as Jupiter, via a spaceship. The interchange of matter as well as information between the Earth as a cell in the universe and another planet as a separate cell is, no doubt, affected by both the structure and the behavior of their respective atmospheres acting as cell membranes. Likewise, the cooperative workings and the exchange of matter between brain cells in the brain universe are influenced by the structure of cell membranes.

Cell membranes, in general, are made of phospholipid molecules, which look like hairpins a few nanometers long (i.e., ten to the power minus nine meters). The head area of a phospholipid resembles the base of the hairpin. Attached to this head is a polar molecule containing a phosphorus atom, which is hydrophilic (i.e., attracted to water). The molecular chains of hydrocarbons at the two ends of the pins are hydrophobic (i.e., water repellent). Since the hydrophilic head area becomes attached to a water molecule, and the hydrophobic tail area is repelled by water molecules, phospholipid molecules form a bi-layer so that the hydrophobic tails are all arranged above and below one plane, yielding a membranous structure. The hydrophilic heads form the interior and the exterior surfaces of the membrane. Described another way, in a double-layered membrane of phospholipids, the hydrophobic tails are all adjacent to each other, sandwiched between the hydrophilic heads. This is the simple structure of the cell membrane. (See Figure 9.) The thickness of this lipid bi-layer is, at most, twice the length of each lipid molecule, which is known to be about three nanometers long, so it can hardly be seen by an electron microscope. However, the structure of a cell membrane can be verified today by means of X-ray or **neutron** diffraction analyses. Moreover, a method of artificially producing a lipid bi-layer has been established by a chemist named Irving Langmuir.

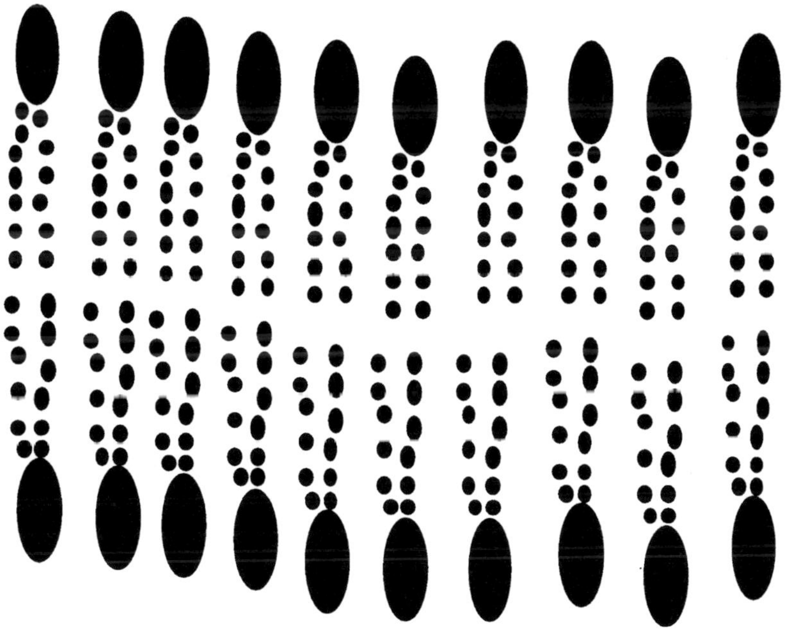

Figure 9. A sectioned diagram of a phospholipid bi-layer. In a double-layered membrane of phospholipids, the hydrophobic tails are all adjacent to each other, sandwiched between the hydrophilic heads.

The cell membrane consisting of a double-layered phospholipid does not have a fixed, static structure; it has a fluid structure that allows phospholipid molecules to move freely in the membrane plane. The geometrical form of the membrane itself can change freely, of course. The membrane can be thought of as a two-dimensional fluid like the atmosphere seen on a global scale. It may be illustrative to think of the membrane as a shallow sea. Nevertheless, this picture of a shallow sea representing the fluidity and the malleability of the phospholipid bi-layer hardly seems to capture the actual character of the cell membrane. True, it is fluid. True, it has functional properties, including permeability to oxygen, carbon dioxide, water, and fat-soluble molecules as well as impermeability to large polymers and ions with electric charge. Nevertheless, the phospholipid bi-layer is not sufficient to yield cooperative workings among the cells. Such activities among the cells would require controlled interchange of matter between the cells, but the simple phospholipid bi-layer does not provide an active control mechanism. In reality, the cell membrane is not a pure phospholipid bi-layer. Particular protein molecules, called membrane proteins, are adrift in the sea of the phospholipid bi-layer like icebergs floating in the sea. It is known that there exist various kinds of membrane proteins that perform specific functions, and, to our surprise, they can move quite freely in the sea of the phospholipid bi-layer. Likening the phospholipid bi-layer to the sea covering the Earth's surface, membrane proteins are large and small icebergs floating there, and they are set adrift by billows and swells.

Furthermore, a lipid bi-layer made solely of phospholipid molecules manifests a phase transition as its temperature falls, and its fluidity is known to decline concurrently with the lowering of the temperature. Hence, in the actual cell membrane, the fluidity is maintained by the mixing of phospholipids with cholesterols and glycolipids. On the exterior side of the membrane, there exist glycolipids, which are nothing but phospholipids with their hydrophilic heads replaced by huge oligosaccharides; these glycolipids are supposed to execute specific functions. Distributed in the cytoplasmic side of the cell membrane, that is, the interior side, are many hydrophilic heads with negative electric charge; together with water molecules, they are thought to produce a two-dimensional layer of plasma just beneath the membrane. This extremely thin, liquid ionosphere just beneath the cell membrane is called bioplasma, the perimembranous region, or the organic Debye layer, and it is known to manifest anomalous physical phenomena such as **superconductivity**. We will revisit this interesting topic in one of the following sections.

Section 29
Membrane Proteins and Ion Channels

Thanks to the chemist Irving Langmuir, pure lipid bi-layers can be produced artificially, and their physical characteristics are now known. Among them, a specific feature is essential: water molecules can penetrate the lipid bi-layers, but ions, that is, charged molecules, cannot pass through. A variety of ions—for example, the hydrogen ion, calcium ion, sodium ion, potassium ion, and chloride ion—play central roles in many physicochemical processes taking place in the cytoplasm. The cell membrane works to confine those ions in the cytoplasm. Still more, it steadies the cytoplasm against the fluctuations in ion density caused by changes in the environment surrounding the cell; however, if the cell membrane cut off ions completely, the metabolic reactions taking place in the cytoplasm would quickly cease. Adequate control of the supply and discharge of ions between the interior and the exterior of the cell is necessary to maintain the physicochemical reactions for energy generation as well as for metabolism in the cytoplasm. The latest research reveals that the flow of calcium ions inside the cell is a fundamental aspect of cell functioning, and the undulatory propagation of calcium ions inside as well as between neuroglia cells plays an important role in realizing the overall workings of the brain.

Then, how and under what conditions are certain ions sometimes allowed to permeate the cell membranes? The existence of membrane proteins is the key. Among the many proteins embedded in the phospholipid bi-layer and distributed over the whole cell membrane, those piercing the bi-layer selectively transport ions, saccharides, and amino acids in and out through the membrane. They are called transmembrane proteins, membrane transport proteins, or simply carriers, and they seem to essentially determine both the structure and the function of the membrane. Brain functioning is thought to arise from cooperative workings among the brain cells, so the cell membrane itself plays a more fundamental role than the cytoplasm. That is, the activity of those carriers in the cell membrane is instrumental to brain functioning. Although the activities of many varieties of membrane transport proteins have not been completely catalogued, we do know that two kinds of carriers transport ions. One is a protein molecule with a hollow core. This protein molecule is called an ion channel. Its geometric structure resembles an Italian pasta, macaroni. The other carrier has a considerably more sophisticated, solid structure. This protein molecule is called an ion pump.

Ion channels have hollow cores that serve as tunnels through which ions and small molecules pass, and some of them have gates that close the tunnels so ions and small molecules cannot always pass through. Two different types of membrane molecules are known to be typical ion channels: one opens the gate by receiving a certain molecule and closes the gate otherwise; the other opens the gate by sensing the decrease of electric potential difference between the interior and the exterior of the membrane surface. Embedded in the membranes of neurons and glia are many ion channels, which form a huge sophisticated network of ion currents inside the brain. Consciousness and memory as functions of the brain might arise from the cooperative functioning of brain cells, such as neurons and glia, while the nature of these phenomena might be revealed in the dynamical network of ion currents spread over the whole brain. We should understand the roles of ion channels embedded in the cell membrane before proceeding to the investigation of higher level brain processes.

Among various membrane transport proteins, a specific one called sodium-potassium ATPase (adenosine triphosphate-ase) plays the most fundamental role. In general, transmembrane proteins take energy from ATP molecules by hydrolysis and use it to produce conformational changes of the molecules as well as to exert mechanical forces. More specifically, the sodium-potassium ATPase produces a sequence of complex conformations of ATP molecules, and it works as an ion pump to expel three sodium ions to the outside and to draw two potassium ions to the inside of the cell membrane. When such ion pumps are functioning, there are more potassium ions inside and more sodium ions outside the cell. As a positively charged element is transported from the inside to the outside during one pumping cycle, the outer surface of the membrane becomes positively charged and the inner surface negatively charged. The difference between the ion concentrations inside and outside the cell membrane may be indispensable to brain activity, considering that most of the energy produced by glucose catabolism and stored in ATPs within the cytoplasms of neurons and glia may be consumed by those ion pumps. A cell membrane is said to be polarized if its exterior and interior surfaces become positively and negatively charged, respectively. The propagation of neural signals is realized in neurons in terms of polarized membranes. Let us now examine more closely the polarization process that realizes the transmission of neural signals.

Neurons exist both in the central nervous system and in the peripheral nervous system. They serve as pathways to transmit neural signals. As for the form, the membrane of a neuron projects outward in some areas and manifests ramification; that is, the projections increase the area of the membrane surface. Those arborized projections are called *dendrites*. In the central nervous system, especially in the cerebral cortex, the majority of neurons have only dendrites, but in the peripheral nervous system almost every neuron has a longer projection called an *axon*. As is easily imagined from this fact, the axon is the pathway through which neural signals are transmitted. Even though it projects far beyond the main body of the cell, the axon is still a part of the neuron, and it manifests a spatial structure in which cytoplasm is enclosed by a membrane. Because the axon has a long and narrow form, electrical signals can propagate along the cell membrane of the axon by means of a molecular mechanism, described as follows:

Among the membrane proteins embedded in the cell membrane, the sodium-potassium ATPase plays the role of an ion pump to maintain the polarization of the cell membrane. However, the existence of such ion pumps results simply in a permanent charge-distribution difference between the interior surface and the exterior surface of the cell membrane; by themselves, the ion pumps could not perform any important function. Ion channels equipped with gates become essential here. The main part of a single neuron, including the dendrites but excluding the axon, is called the cell body. There are many ion channels equipped with gates distributed over the membrane of the cell body. Some of them open their gates for a while if they receive specific molecules called neurotransmitters. If such a receptor-sensitive ion channel opens its gate, the excessive positive charge of sodium ions outside the cell flows into the cell through the open tunnel and temporarily cancels out the charge-distribution difference between the interior surface and the exterior surface of the polarized membrane near the channel. This event is called depolarization since the electric-potential difference between the inner and outer surfaces of the membrane diminishes. Of course, the electric-potential difference will recover in awhile as the gate closes, and the depolarization of the membrane disappears.

What will happen, however, if there are some electric-potential-sensitive ion channels near this receptor-sensitive ion channel that open their gates in response to depolarization as a result of a decrease in the electric-potential difference between the interior and the exterior surfaces of the membrane? Because depolarization diminishes the electric-potential difference between the inner and outer surfaces of the membrane, those neighboring electric-potential-sensitive ion channels open their gates, so sodium ions flow through them into the cell, too. Consequently, the depolarized region of the membrane spreads out, and a fluctuation of electric potential spreads across the surface of the cell body. If no electric-potential-sensitive ion channels are distributed near the depolarized region of the membrane, sodium ions that flow inside the membrane diffuse as time passes, and the depolarized region extends gradually and weakens the degree of depolarization. More simply put, the spread of the depolarized region due to the diffusion of ions causes a decrease in the degree of depolarization.

The cell membrane is depolarized in the vicinity of receptor-sensitive ion channels distributed over the surface of the cell body when they receive the neurotransmitters, and the depolarization spreads over the whole surface of the cell body. If a number of ion channels happen to receive neurotransmitters at about the same time, they generate depolarizations in their own vicinities that spread equally across the whole surface of the cell body. Then, the electric-potential difference between the interior and exterior surfaces of the cell body membrane manifests a fluctuation called the generator potential, which is the sum of those different depolarizations. In other words, the spatiotemporal change in the distribution of the membrane-electric-potential difference on the surface of the cell body of a neuron is triggered by many receptor-sensitive ion channels receiving the neurotransmitters, and that change is controlled by many electric-potential-sensitive ion channels.

Section 30
The Axon and Neural Impulse

Since fluctuating differences in the ion concentrations on the interior and exterior surfaces of the cell body membrane are triggered by the material environment of the extracellular region, that is, the existence of neurotransmitters, brain function may be seen to arise from cooperative activity among neurons exchanging neurotransmitters. Indeed, this point of view has been overestimated in modern, twentieth century brain science and has given rise to the so-called neuron doctrine, which is still upheld by most brain scientists today. This might be due, in part, to the fact that the fluctuations of the electric-potential difference between the interior surface and the exterior surface of the cell membrane, induced by the opening of the gate of receptor-sensitive sodium ion channels (depolarization), is on the order of tens of millivolts and, thus, easily observed even within brain tissue.

In the neuron doctrine, the axon of the neuron plays an important role. As we have seen, the spatiotemporal fluctuations of the electric-potential difference between the two surfaces of the cell body membrane result from events in which many ion channels distributed on the membrane receive neurotransmitter molecules. Then, how might this disturbance in the distribution of the electric-potential difference produced in a single neuron by some external action of neurotransmitters work on other neurons? This effect is thought to be a function of the axon. Since the axon is in the form of a long projection, it may be regarded as a long and narrow tubular membrane enfolding cytoplasm. On the tubular membrane, electric-potential-sensitive ion channels are distributed adjacently. If the electric-potential difference between the inner and outer surfaces of the cell body membrane at the root of the axon is diminished, causing the gates of ion channels to open there, inflowing sodium ions induce depolarization. Diminished electric-potential difference due to this depolarization extends to the neighboring electric-potential-sensitive ion channel and, in turn, opens its gate. Repeating this process, depolarization of the membrane propagates along the axon. A membrane-electric-potential difference of tens of millivolts shifts its position along the tubular membrane of the axon; this membrane-electric-potential difference is called a *neural impulse*.

The neuron doctrine proposes that this unique constituent, the neural impulse, is responsible for the advanced functions of the brain, namely consciousness and memory. The following factors are thought to constitute the advanced functions of the brain:

- Neurotransmitters adhering to the surface of the cell body of a neuron trigger a neural impulse, that is, a pulse-like propagation of membrane-electric-potential difference of a magnitude sufficiently large for direct observation by microelectrodes.

- There are electric-potential-sensitive calcium ion channels at the terminal knob of the axon, and their gates open when the neural impulse arrives at the terminal knob.

- Calcium ions flow into cytoplasm through the calcium ion channels so that endoplasmic reticula containing neurotransmitters are released to the outside by merging into the membrane of the terminal knob.

- The released neurotransmitters adhere to the surface of the cell body of another neuron adjacent to the terminal knob and trigger a neural impulse.

By combining and repeating this process, neurons are believed to manifest a sequence of "advanced functions," as follows:

- The neuron converts several incoming signals of neurotransmitters into electric-potential signals via ion channels distributed on the membrane surface of its cell body.

- The neuron combines those electric-potential signals at the root of the axon.

- The neuron produces a neural impulse there if the combined membrane-electric-potential difference becomes smaller than a certain threshold value.

- The neuron shifts the neural impulse toward the terminal knob of the axon.

- The neuron releases the neurotransmitters to the exterior of the membrane of the terminal knob.

In the peripheral nervous system, such a function of the neuron is, indeed, put to good use. It is an obvious fact that each bodily tissue establishes an information-transmission pathway in terms of neural impulses and neurotransmitters. However, it is known that in the central nervous system, especially in the cerebral cortex, the majority of neurons have no axons and cannot transmit neural impulses. A recent study seems to reveal that the highly advanced workings of the cerebral cortex are better understood by certain microscopic physical processes taking place in the sophisticated network of dendrites of neurons without axons, that is, in the dendritic network (Pribram 1991). Nevertheless, the neuron doctrine still survives, and consciousness and memory, as functions of the brain, are believed to arise from the transmission pattern of neural impulses in the neural network of axons of neurons.

Programs simulating the function of neural networks have been developed for the ordinary digital electronic computer, and it has become popular to talk about the "neurocomputer" in computer engineering. The neurocomputer simulating the function of neural networks is receiving media coverage as an epoch-making computer architecture. This "neurocomputer," however, might be overestimated since the essential aspect of the human brain comes from networks other than the neural networks.

Section 31
The Neural Network

The generation and propagation of a membrane-electric-potential difference of the order of tens of millivolts is called a neural impulse. It is also called an action potential of a neuron. It is often said that the neuron "fires" when it generates an action potential. As we have already seen in the case of a neuron with an axon, the membrane-electric-potential difference fluctuates as a result of depolarizations caused by neurotransmitters adhering to the surface of the cell body, and the neuron triggers the action potential along the axon if the total electric-potential difference at the root of the axon becomes smaller than a certain fixed value, that is, the threshold value. This action potential propagates along the axon as a neural impulse and works at the terminal knob to release neurotransmitter molecules. Therefore, a neuron equipped with an axon functions as a device that controls incoming and outgoing chemicals gated by neurotransmitters.

What will happen then if the terminal knob of the axon of one neuron is placed adjacent to the cell body of another neuron? Of course, generally there will be a considerable number of terminal knobs of axons adjacent to the cell body of each neuron. The totality of neurons thus connected to each other by axons is the neural network, which may be regarded as an electrochemical circuit composed of a great number (about ten to the tenth power) of circuit devices of the multi-input and single-output type with threshold-majority-decision logic functioning. In other words, each neuron as a circuit device of the neural network will produce a single chemical output signal only when the sum of several inputs exceeds a certain threshold value, but it will manifest no reaction at all if this condition is not met. As far as the electric-potential difference between the interior and exterior surfaces of the neuronal membranes is concerned, the neural network may be understood merely as an electric circuit in which about ten to the tenth power multi-input, single-output, threshold-majority-decision switching devices are wired together by axons. Note that the brain is recursively organized—systems embedded within systems embedded within systems—so that the ten to the tenth network is the most complex level of recursive organization and a hyperneuron with a few thousand nodes is the least complex level (Globus 1995).

The neural network is not a simple circuit with a fixed structure but manifests freedom in its circuit wiring for two reasons. First, the neurons are surrounded by glia and exhibit amoeba-like movements. This is especially evident in the terminal knobs of the axons, which move around continually and thrust the glia aside as they search for suitable connection positions. Second, the receptor-sensitive ion channels embedded in the region of the cell body membrane directly adjacent to a terminal knob of the axon of another neuron can float in the membrane and change their positions. The region including the terminal knob and the adjacent area is called a chemical synapse or simply a *synapse*. If there are no receptor-sensitive ion channels in the vicinity of the synapse, depolarization will not take place. Even if neurotransmitter molecules are released from the membrane of the terminal knob, depolarization cannot occur without receptor sensitive ion channels in the vicinity of the synapse. Furthermore, even when many ion channels are distributed there, some of them might work to suppress the depolarization, that is, to increase the difference in the charge distribution between the interior and exterior surfaces of the membrane. They are called inhibitory ion channels, and they are activated by several kinds of neurotransmitters different from those that activate the ordinary depolarization-promoting, receptor-sensitive ion channels. Therefore, even if many ion channels exist in the vicinity of a synapse and many neurotransmitter molecules are released there, depolarization due to the action of ordinary receptor-sensitive ion channels can be canceled out by the existence of inhibitory ion channels.

The neural network, in which numerous neurons are wired together by axons, manifests freedom in the connectivity among the synapses. This freedom is revealed in the conformational change of the synapse and in effective change of synapse functioning. The conformational change is the result of the amoeba-like movement of the axon, and the functional change is the result of the distribution pattern and the species of receptor-sensitive ion channels in the synapse. This freedom is called *plasticity* of the synapse. If we imitate the neural network by regarding it as an electric circuit and fabricate an imitation with standard electronic device technology or perform a computer simulation with ordinary digital computers, it is necessary to assume certain weighting factors for the input contacts of each multi-input, single-output, threshold-majority-decision switching device reflecting the plasticity of the actual synapse contact. Some empirical or phenomenological weighting rules are known, but the mechanism that maintains the plasticity of an actual synapse is not known. Contemporary brain science has identified neither the control mechanism nor the factors that control the distribution of the receptor-sensitive ion channels on the cell membrane, which determine the plasticity of each chemical synapse. Indeed, those discoveries might remain far beyond the scope of contemporary brain science because the detailed microscopic physical phenomena taking place in the cell membrane have hardly been taken into account in brain science. One cannot blame brain scientists for this oversight. Remember that mainstream, modern brain science has been established by medicine, biology, chemistry and psychology. None of these sciences is precisely oriented toward an exploration into imperceptible microphysical phenomena taking place in the cell membrane.

We are apparently approaching a turning point at which we dare to stray from the mainstream of modern brain science in our search for a fundamental understanding of our cryptic universe, the brain. Neurons with axons are, by no means, the majority in the cerebral cortex, which is regarded as the proper site for the emergence of highly advanced mental activity in humans. Neurons without axons and glia are the majority there. Keep in mind that the information exchange between the brain and each bodily tissue is realized by the pathway conduction of neural impulses through the neural network of neurons with axons. Thus, it might be more natural to think of the role of the neural networks in the cerebral cortex as a mere input-and-output interface between the major systems of neurons without axons, as well as the neuroglia cells and bodily tissues. That is, the neural network receives neural impulse signals from bodily tissues and transforms them into other signals to be used by the major systems. Furthermore, the neural network transforms the resulting cooperative activity of the cell assembly of neurons without axons and glia in the major systems into a sequence of neural impulses so that such cooperative activity can govern the bodily tissues.

Today, research in the fields of molecular biology and biophysics has gradually revealed that neurons without axons and glia play the principal roles in the fundamental processes of the brain. *Therefore, the neuron doctrine or the neural network doctrine might be on the decline. One can safely say that memory and consciousness as actions of the brain come not from the neural network of neurons with axons but from certain microscopic physical phenomena taking place in the cell assembly of neurons without axons and glia, which previously were thought to simply maintain and feed neurons with axons.*

Can you imagine what might have happened if one had proposed such a theory in the 1950s and contested the neuron doctrine maintained by mainstream brain science at that time? He or she might have met the same kind of resistance that Giordano Bruno and Galileo Galilei did when they opposed the geocentric doctrine. The prevailing scientific community had accepted the neuron doctrine, which advocated that the neural network of neurons with axons is the fundamental ground of highly advanced brain activity. The neural impulse was considered to be the sole basis for information transactions in the brain. Karl Pribram challenged certain tenets of the neuron doctrine of modern brain science when he focused not on the neural impulse but on the graded change of membrane electric potential. Pribram redefined the states of the neural network to emphasize the graded synaptic changes of membrane electric potential.

Section 32
Grandfather Cells

Karl Lashley was a neuropsychologist in the United States who pursued the trace of memory, that is, the engram, and advanced the theory of nonlocality of memory in the cerebral cortex. He invited a young brain surgeon from Austria, Karl Pribram, to perform experiments on animals in which he created local lesions in their brains and recorded the behavioral effects. The two Karls engaged in exhaustive experimentation to find evidence of memory localization in the brain, but the results of their work offered no support for the existence of the engram. Finally, Lashley advanced the doctrine of memory nonlocalization. Lashley died during the second half of the twentieth century when Penfield's doctrine of memory localization and the neuron doctrine had gained a stronghold within the community of brain scientists. Undaunted, Pribram continued his research along the lines that he had pursued with Lashley and completed the local lesion experiments on the brain. Then he initiated a second type of experimentation in which he recorded the action potentials of certain neurons simultaneously by putting fine microelectrodes into the brain tissue of animals. He expected to obtain without damaging the brain more precise results supporting the nonlocal theory of memory than had been acquired through the local lesion experiments.

Contrary to his expectations, the results of this new experiment seemed to support not only the local theory of memory but also the neuron doctrine. Investigating first the relationship between a visual stimulus and the fluctuations of action potentials of neurons in the visual area, Pribram found neurons that fired in response to a specific stimulus. The existence of such neurons in the visual area had been discovered already by David Hubel and Torsten Wiesel of Harvard University in 1963; nevertheless, he obtained the most precise experimental data on the mechanism of visual perception in the cerebral cortex. After Hubel and Wiesel had demonstrated the existence of such neurons, many scientists pointed to that discovery as evidence of the locality of memory and used it to validate the local theory of memory, which, in turn, promoted the neuron doctrine. But is that interpretation correct? Let us investigate further.

Experiments by many brain physiologists—including Hubel, Wiesel, and Pribram—have revealed that there are a number of neurons with specific characteristics in the visual area of the cerebral cortex, which receive visual signal impulses from the central retina. For example, a certain neuron fires if, and only if, a right oblique line comes into sight. Another one fires for a line at a different oblique angle. Furthermore, there is a neuron that fires only when a line moves in a specific direction in the visual field. Another neuron, also found in the visual association area, fires only when a rotating body comes into view. Those neurons in the cerebral cortex that fire in response to specific visual stimuli are called feature detectors or, metaphorically, "grandfather cells." The metaphorical name suggests the exaggerated situation in which a neuron fires only when a person sees the set of features characteristic of his or her grandfather's face.

This question naturally arises: Why do the neurons called grandfather cells only respond to particular stimuli? Here's the simplest answer: At the time the neuron is produced in the neural tube, it is differentiated to perform a specialized function. That naive notion gives victory to the neuron doctrine, which claims that both brain function and memory are localized in terms of neurons. According to this view, the grandfather cell is differentiated as such from the early stages of its development.

Unlike many other people, Pribram never accepted such a simple idea. The keen intuition and deep insight he gleaned from his work with Lashley, together with his excellent experimental skills, were put to good use to elucidate a higher order of existence lying concealed behind the grandfather cell. Pribram used a computer system to record and analyze experimental data. Specifically, he recorded the firing pattern of a grandfather cell in the visual area of the cerebral cortex and the membrane electric potential of the neighboring neurons as well as the neural impulse signals coming from the central retina through a neuron of the lateral geniculate body. Here are the results: If the grandfather cell fires in response to a series of neural impulses transferred from the central retina, the grandfather cell is not activated independently of the neighboring neurons; on the contrary, the membrane electric potential of the grandfather cell and that of the neighboring neurons manifest strong mutual correlation, though only the grandfather cell fires. What on earth could that mean?

Even though only the grandfather cell fires in response to a stimulus, the neighboring neurons also manifest meaningful reactions. Not only does the grandfather cell respond to the visual stimuli, the neighboring neurons also react in a subthreshold region; that is, they manifest a correlated graded synaptic change of membrane electric potential. Pribram offered the following explanation for these experimental results. In the visual area of the cerebral cortex, the dendrites of a number of neurons, including the grandfather cell in question, become entangled in a sophisticated type of net and form a dendritic network. As a result of cooperative activity within the dendritic network, one of the participating neurons fires when a person sees the face of his or her grandfather.

Extending Pribram's account of visual perception to the fundamental processes of memory and consciousness in the cerebral cortex, we expect that the more essential role is played there, not by the neural network of neurons with axons, but by the dendritic network formed by the dendrites of neurons with and without axons, which surround the neural network like a dense tangle of ivy covering a tree. Since the dendrites of neurons with axons in the neural network belong to certain dendritic networks, the result of some cooperative physical process in the dendritic network might be found in the activity of a certain neuron with an axon. Such dendritic networks are densely interwoven in the outer layer of the cerebral cortex, and they extend behind the first layer of neurons. The neurons in that first layer initially receive the neural impulse signals from bodily tissue and from the neurons with axons forming the neural network connected to this first layer of neurons. They seem to produce a microscopic mechanism for the realization of highly advanced brain activity. Indeed, inspired by his own experimental results, Pribram has paid particular attention to cooperative physical phenomena in the dendritic networks densely distributed over the cerebral cortex. He has looked there for the principle governing highly advanced visual processing and the mechanism of memory. The dendritic network seems to provide a microscopic structure in the cerebral cortex, which can be pursued through a physiological method. Hence, it may be worthwhile to look at the structure of a typical dendritic network.

Section 33
The Dendritic Networks

A dendritic network is a microscopic structure composed of the dendrites of numerous neurons with or without axons entangled with each other in a highly sophisticated manner. The dendritic network is surrounded, of course, by neuroglia cells, so they also might participate in the cooperative physical phenomena within the dendritic network. We will revisit such a possibility, but for the moment we shall restrict our discussion to neurons. Dendrites are nothing but tree-like projections of the cell-body membrane. The longer ones are difficult to distinguish from the axon, in either appearance or function. Observation by an electron microscope reveals many spines on the surface of the dendrites; these are called dendritic spines or simply spines. In 1959 Edward Gray found that chemical synapses form on the tips of the spines. Most of the spines in the cerebral cortex are known to have a mushroom-like form about one micron in length and zero-point-one micron in diameter.

While the dendrites of numerous neurons form the dendritic network by entangling with each other in a highly sophisticated fashion, the manner in which they are joined is not restricted to the formation of chemical synapses at the tips of the dendritic spines. The majority of connective structures are actually not chemical synapses. There are electric synapses, also called ephapses, as well as tight junctions or gap junctions. The ephapse is a simple structure in which two dendrites just barely contact each other, and the membrane electric potential arising from the charge-distribution difference between the inner and outer surfaces of the membrane of one dendrite leaks out to the other dendrite. The tight junction or gap junction is a slight gap between adjacent dendritic membranes belonging to two different neurons; it is created when groups of small membrane proteins form a hole about one nanometer in diameter on each of those dendritic membranes.

By investigating the spatial structure of neurons in the cerebral cortex, we find that many developed dendritic networks overlay the neural network composed of neurons with axons. Especially in the outer layer, the principal microscopic structure is the dendritic network. As we have already seen, neurons with axons only provide pathways for neural impulses. The neural network of such neurons is nothing more than an interface device that performs several functions:

- The neural network transforms the result of cooperative physical phenomena taking place in the microscopic structure of the cerebral cortex into the flow signal of the neural impulse.

- The neural network transmits the flow signal of the neural impulse to each part of the body.

- The neural network receives the stimulus signal of the neural impulses transmitted back from each sense organ.

With that in mind, it is hard to believe that consciousness and memory arise from the activity of the neural network. It might control and transmit information governing, lower order processes, such as reflex movement, but never the higher order processes that maintain consciousness. The activity of the neural network would be limited to psyche, a shabby, unintelligent human soul that does not participate in consciousness. Even in mainstream cerebral physiology, the higher order process is thought to be realized by the cooperative activity of dendritic networks, which are fine structures overlaying the neural network of neurons with axons.

Pribram pointed out that highly advanced visual processing is realized by microscopic physical phenomena taking place in the dendritic network lying behind the neural network of the visual area, and only as a net result does the grandfather cell fire. Thus, the idea has been gradually accepted that the dendritic network plays an important role in the production of the highly advanced functions of the cerebral cortex, namely memory and consciousness.

Yoshiya Shinagawa, for example, paid considerable attention to the role of dendritic networks in relation to brain waves. In the chapter on the classical theory of the brain, we noted a few known facts about the origin and the mechanism of brain waves, that is, the temporal fluctuations of electric-potential distribution over the scalp, as measured by electrodes. In the golden days of the neuron doctrine, human consciousness was seen as arising from the action of the neural network of neurons with axons in the cerebral cortex, and brain waves were regarded as nothing but the accumulation of the action potentials of all the neurons in the neural network. That belief is no longer widely held. Indeed, the contributions from many neurons cancel each other, and their accumulation vanishes. Brain waves are not generated by the propagation of neural impulses in the neural network. As we pointed out earlier, in order for brain waves to be observed in the scalp region, there must be completely synchronized fluctuations of electric potential spanning at least six square centimeters of the outer layer of the cerebral cortex. Shinagawa speculated that brain waves arise from certain cooperative physical phenomena in the dendritic network because that appears to be the most plausible microscopic and global structure in the outer layer of the cerebral cortex to satisfy a global synchronous condition. If brain waves are produced by the dendritic networks of the outer layer of the cerebral cortex, it seems quite possible that we will find the mechanism that realizes the highly advanced functions of consciousness and memory in the dendritic networks.

Thus, brain functions that account for consciousness and memory are now thought to arise from cooperative physical phenomena that take place in the dendritic networks of the cerebral cortex. Then, what kind of physical phenomena are they? Pribram has suggested that electric charges on the membrane of the spine head do not flow into the dendritic membrane because of the high electrical resistance of the spine. That is, each spine head might maintain electric-charge distribution and, as a result, an **electric dipole** might be formed there. He studied the cooperative physical phenomenon arising from electrical interaction between the dynamical system of electric dipoles on the spine heads and the dynamical system of the membrane electric potential of the dendrite. The mechanism of the highly advanced functions of the cerebral cortex, namely, consciousness and memory, can be found neither in the neural impulses running around the neural network of neurons with axons nor in synapse plasticity, but in the dynamical interaction between the system of electric dipoles of each spine head and the dendritic membrane electric potential. Today, *dynamics* might be the key word for a better understanding of brain function.

Chapter 6
Dynamical Theory of the Brain

Section 34
Physics and Dynamics

We have explored the material structure of the brain in our search for the mechanism of **consciousness** and **memory**. Modern cerebral physiology has developed through the use of the scanning electron microscope, and the dendritic network has come to the front line of investigation. The prevailing question now is whether the dendritic network of neurons without axons as well as the cell assembly of glia, which were overlooked in the neuron doctrine, perform the fundamental processes in memory and consciousness, as emphasized by Pribram (1991). This, at least, seems clear: the answer to this question might not be found within the realm of modern cerebral physiology. Brain functioning and the mechanism of consciousness were peripheral concerns in the Cartesian model of nature and cosmology. Within that mechanistic framework, brain function became a subject of analytical investigation in physiology, while concepts in physics that provided a fundamental view of nature became estranged from brain science. Now, however, physical science and brain science seem to be converging in a shared view of nature. A dynamical point of view, quite familiar in physics, has been remote from brain science until now.

In physics a wide range of natural phenomena that occur in our world have been investigated from the universal and fundamental point of view of dynamics. Although each phenomenon has its own spatial and temporal scale and a distinct law governing it, it can also be understood in terms of *dynamics*, a common physical concept defined as "controlled changes." Thus, every natural phenomenon in our world is subject to controlled changes and is, therefore, dynamical. The particular law governing a newly discovered phenomenon might not be known initially, but the dynamics underlying the phenomenon can be described without recourse to the underlying physical law. Put another way, the dynamics can be described phenomenologically, without reference to a governing principle.

For example, the dynamics of the whirlpool of our galaxy can be the same as that of a vortex of water in a washbasin, though their spatiotemporal scales and particular laws are quite different. The dynamics of a drunkard tottering around *Ginza* in Tokyo resembles the zigzag movement of a small particle of pollen in water as botanist Robert Brown observed. When physicists discovered plasma—the fourth phase of matter following solid, liquid and gas—they noted that its positively and negatively charged constituents gather and scatter subject to electromagnetic interaction; its dynamics, resembling that of a living cell, evoked the name "plasma." (Recall the cytoplasm in a cell.)

Dynamics, of course, can be described by starting from first principles, as is the approach in **quantum mechanics** and **quantum field theory**, but one can also assume a certain phenomenological principle, as is the case in many other research fields of physics. Thermodynamics, hydrodynamics, aerodynamics, electrodynamics, and magnetohydrodynamics are all built on phenomenologically-based principles. Today when developing a physical theory to understand a certain natural phenomenon, one is not expected to identify the underlying principle, such as the fundamental law of **quantum theory**, but to provide a dynamical description that points to a peculiar phenomenological principle. For example, the mechanism of James Watt's steam engine can be well understood without a thorough knowledge of the molecular structure of steam; one need only apply the phenomenological principle of thermodynamics, such as the first law of thermodynamics (Mayer and Joule), the second law (Clausius and Lord Kelvin), and the third law (Nernst and **Planck**). Regarding steam as gas, one can apply the kinetic-molecular theory of gases, which was derived from the first principle of **classical mechanics** or quantum mechanics, to understand the mechanism of the steam engine.

Our final aim in the present volume is to present a physical theory to explain consciousness as a natural phenomenon. There will be two different approaches: one is to find a phenomenological principle, and the other is to start from first principles. Before proceeding to an examination of first principles in quantum mechanics and quantum field theory in the next chapter, we shall present a dynamical and phenomenological description of the material structure of dendritic networks, as observed through experiments, and discuss the phenomenological principle at work there.

Section 35
Brain Science and Cybernetics

The natural phenomena in our world are generally divided into two groups:

living and non-living. Physics deals primarily with the latter; that is, the fundamental principles governing non-living matter are the main concern of physics. Recall classical mechanics, electromagnetism, thermodynamics, fluid dynamics, and so on. The concept of dynamics, however, can be applied equally to living and non-living entities, for both of them undergo controlled changes. The dynamical point of view has been fully employed in physics to account for the phenomena of non-living matter, but it has been outside the scope of biology, which is the study of life phenomena. This was also the case in brain science, but an ingenious mathematician helped to bring dynamics into brain science. The mathematician was Norbert Wiener, who introduced the notion of *cybernetics*, which pertains to the dynamics of both life phenomena and artificial machine systems and emphasizes control and communication (Wiener 1948). The term cybernetics, chosen from the Greek expression for "steersman" in 1947, highlighted the importance of feedback and feedforward mechanisms in the controlled changes of life phenomena. Since then cybernetics has been discussed mainly in reference to artificial brains or artificial intelligence. Derived from cybernetics and organism, the term *cyborg* came into being.

Wiener inadvertently opened a new frontier in modern brain science by laying the track for an exploration into the brain from a dynamical perspective, though cybernetics itself does not fit into the mainstream of brain science. The mainstream has always adhered to the physiological approach guided by **molecular biology**. Cybernetics has been regarded as a mere theoretical framework to provide simplified models of the functioning of the brain. For example, the neural network of neurons with axons falls under cybernetics if it is seen as a kind of a sophisticated electric circuit. Then the controlled change, that is, the cybernetics, of the neural network model can be investigated as the dynamics of a specific electrical circuit consisting of numerous switching devices. Thus, the concept of dynamics first entered into modern brain science through cybernetics and neural network models.

Section 36
A Neural Network Model

Let us look at a standard way of describing cybernetics by taking the dynamics of a neural network model as an example. Afterward, we will conduct a detailed investigation of the dynamics of the dendritic network. We shall consider a large scale integration of neurons with axons in a typical cell assembly of the cerebral cortex. Each neuron is connected to many others by chemical synapses, and the totality of neurons may be regarded as a cybernetic system with numerous **degrees of freedom**, up to the order of ten to the tenth power. This system exhibits cybernetics in its complex interactions involving controlled propagation of pulse-like electrical signals, called neural impulses, along the axons; the feedforward and feedback circuitry control, due to recurrent input, is assumed to account for synapse plasticity.

Although there exist finer structures behind the neural network, such as the dendritic network, we are ignoring them for the moment simply because here we are interested in exposing the cybernetics of the neural network. (A more refined formulation including the dendritic network will be given in the following sections.) Neurons belonging to the neural network are called front neurons or visible neurons if they are connected to neurons that do not belong to the neural network in question. The remainder are called hidden neurons.

Let us look at the simplest degrees of freedom, that is, the change in the dynamical parameter of each neuron in the neural network. As shown in the preceding chapter, the action of a neuron can be described by the cell-body-membrane-electric-potential difference relative to a threshold value at the root of the axon. That is, the neuron can be regarded as a switching device of electric circuits, and the switching on or off is determined by comparing the membrane-electric-potential difference at the root of the axon with a given threshold value. A neuron fires and generates a neural impulse along its axon when the membrane-electric-potential difference at the root of its axon becomes smaller than a threshold value, which is a given value that indicates the existence of a hurdle to leap over in order to realize a controlled change in the degree of freedom.

We choose this membrane-electric-potential difference relative to the threshold value as the simplest degree of freedom of a neuron in the neural network, and we call it a generator potential. Thus, a neuron fires only when its generator potential appears negative, that is, when the membrane-electric-potential difference at the root of the axon becomes smaller than the threshold value. In terms of the generator potential of each neuron in the neural network, its dynamics, that is, its controlled change, can be well described by the following equation:

{Increment of}[Generator Potential]
=[Electrochemical Interaction][Time Interval]+[Fluctuation]

In symbolic notation, this becomes

$$\Delta X = \Phi \Delta t + \Xi$$

where ΔX denotes an increment of generator potential in a time interval Δt, Φ denotes electrochemical interaction, and Ξ the fluctuation. In simpler language, the above equation means this: *Each increment of the generator potential of each neuron equals the purely electrochemical interaction between the neuron in question and others through chemical synapses times the time interval plus a fluctuation caused by a combination of numerous uncontrollable factors.*

The purely electrochemical interaction is called a drift, and it depends on the generator potentials of all the neurons in the neural network. The drift is nothing but the speed of the change of the generator potential obtained by comparing the best estimate of the unknown value of the generator potential in the future and its given present value. The uncertainty caused by a combination of numerous, uncontrollable factors is simply called fluctuation. The mathematical theory of **probability** informs us that such a combination becomes a **random variable** subject to a **Gaussian probability law**, regardless of the details of the uncontrollable factors, provided that the number of the uncontrollable factors is large. This fact, called the central limit theorem or the law of large number, ensures the fluctuation to be a **Gaussian random variable** with vanishing mean value and definite variance.

To investigate the detailed characteristics of cybernetic systems, Wiener developed a general mathematical theory to treat the following equation (Wiener 1948):

{Increment of} [Generator Potential]
=[Drift][Time Interval]+[Fluctuation]

Or symbolically,

$$dX(t) = A(t)dt + dB(t)$$

where $dX(t)$ is an increment of generator potential in an infinitesimal time interval dt, $A(t)$ is a drift, $dB(t)$ is a Gaussian random variable standing for the fluctuation and t is a time parameter. The above equation means that *the increment of the generator potential equals the drift times the time interval plus the fluctuation.*

This equation, later called a stochastic differential equation, was fully developed by Kiyoshi Itô (Nelson 1967; 1985). Recall that the drift is nothing but the speed of the change of the generator potential obtained by comparing the best estimate of the unknown value of the generator potential in the future and its given value at present. Therefore, the stochastic differential equation provides a mathematical tool for reproducing the controlled change of the generator potentials of all the neurons by estimating as best one can their successive increments, step-by-step, starting from some initial values and integrating them through a method called the Wiener integral or the stochastic integral (Nelson 1967; 1985). In other words, the cybernetics of the neural network can be described by the stochastic differential equation provided that we know both of the following:

- the functional form of the drift depending on the generator potentials of all the neurons in the neural network

  and

- the fluctuation subject to the Gaussian probability law.

As for the fluctuation, we need to obtain the variance parameter, which might be common to all the neurons. The data from recent experiments with microelectrodes can be used to fix this statistical parameter. Unlike the relative ease with which we can find the fluctuation included in the stochastic differential equation, it is somewhat difficult to determine the functional form of the drift of even a single neuron. The reason is that it depends not only on the generator potential of the neuron in question but also on those of the other neurons in the neural network, which could amount to ten to the tenth power in number. Thus, the principal goal in the cybernetics of the neural network is to obtain a mechanism of control over the sophisticated synapse connectivity between neurons reflected in the drifts.

Of course, we have already seen that such a control mechanism has not been found in modern physiological theories about the brain. In cybernetics, however, we can use a model control mechanism extracted phenomenologically from the experimental data in a simple case. The connectionist model might be the most popular model in modern brain science. There, each drift is chosen as a piece-wise linear function of the generator potentials of all the neurons so that it gives the gradient of steepest descent of a multidimensional potential designating the configuration energy or Hopfield's computational energy of the mutual connectivity among neurons in the neural network (Yasue et al. 1988). (Piece-wise linear means locally linear. Imagine a zigzag line. It is not linear as a whole but linear in each piece. The details of this connectionist model will be explained in the next section.) The drift is fixed in such a way that it optimizes a certain quantity. That is very important in developing the model of cybernetics. There has been a strong and tacit belief in science that the controlled changes in natural phenomena are optimized. Hence, it is natural to expect some optimization principles running through the formulation of cybernetics. The concept of optimal control is the key concept in the control theory of artificial systems, as it is in cybernetics. In the following sections, we will see why optimization principles are essential in the cybernetics model of brain activity.

Section 37
The Spin-Glass Model

The cybernetics of the neural network can be described by the generator potentials subject to the stochastic differential equation. In this neural network model, all the degrees of freedom seem to suffer from controlled changes due to mutual interactions as well as from uncontrollable fluctuations. This question naturally arises: Can the neural network model realize a mechanism to maintain a memory of some external stimulus? In other words, does the neural network model exhibit a memory storage function?

Let us recall the stochastic differential equation governing the generator potentials of the neural network. It tells us that the increment of each generator potential never vanishes identically, so it can undergo a controlled change forever; therefore, the neural network model seems to fail to realize a memory storage mechanism. Nothing can be maintained when all the degrees of freedom are changing. However, this might not be the case. Let us suppose that the drifts vanish in a peculiar configuration of the values of the generator potentials. We will call that a local equilibrium configuration. There are, in general, as many local equilibrium configurations as there are degrees of freedom, that is, about ten to the tenth power. If the generator potentials happen to take values close to those of a local equilibrium configuration, their increments continue to manifest only Gaussian fluctuations with vanishing mean values. Then, as long as the variance parameter of the common Gaussian fluctuation remains small, it is most probable that the generator potentials will continue to take the same values as the local equilibrium configuration. We can interpret this result to mean that the cybernetics of the neural network maintains the most probable values of the generator potentials close to the values of the local equilibrium configurations of the neural network.

It seems worthwhile to note here that local equilibrium configurations are nothing but local minima of the "configuration energy" of the mutual connectivity among neurons in the neural network. Thus, they are candidates for the representation of a sustainable short-term memory. More simply put, the local equilibrium configurations are thought to be short-term memories. Long-term memory is expected to be sustained by the change of the mutual connectivity among neurons in the neural network, though the mechanism is not yet known.

Let us illustrate a process of imprinting the memory of an external stimulus. In the neural network model, any external stimulus can be thought to forcibly set the values of the generator potentials of the visible neurons. The initial values for the hidden neurons reflect the state of cybernetics of the neural network just before the external stimulus in question registers. Starting from those values of the generator potentials of all the neurons, the neural network undergoes a controlled change governed by the stochastic differential equations. Then, after awhile, the generator potentials happen to take the values close to those of a certain local equilibrium configuration of the neural network, and most probably they will keep the same values until the next external stimulus is transferred. Think of the local equilibrium configuration as a storeroom for the external stimulus in question. The corresponding values of the generator potentials of the hidden neurons will combine with the new values of the visible neurons, which are given by the next external stimulus, and the memory imprinting process will again be triggered.

This memory imprinting process of the neural network model might be well illustrated schematically by introducing a dichotomous random variable for each neuron. Let us think of the inverse sign of each generator potential: It takes the value *-1* whenever the generator potential takes a positive or vanishing value, and *+1* whenever it takes a negative value. We regard this sign as an activity indicator for each neuron in the neural network since it reflects the two distinct values *-1* and *+1* corresponding to the distinct states of a neuron, inactive and active, respectively. Thus, we call this inverse sign the activity indicator of a neuron. The totality of activity indicators of all the neurons can describe the cybernetics of the neural network. Although they suffer from controlled changes as time passes, their values are fixed at either *+1* or *-1*.

We can follow the time change of the activity indicator of each neuron by rewriting the "configuration energy" of the mutual connectivity among neurons in terms of those activity indicators. Taking into account the stochastic differential equation that governs the increment of the generator potential, the activity indicators can be understood to also approach the values of an equilibrium configuration, subject, of course, to the uncontrollable fluctuation. This time, the values of an equilibrium configuration corresponding to a local minimum of the "configuration energy" manifest a distribution in which each value is either *+1* or *-1*.

Let us try a thought experiment. Picture the neurons in the neural network as a mass of neurons with definite volume and surface. Now we rearrange them in such a way that only visible neurons form the surface and hidden neurons are confined inside the mass. The visible neurons literally become visible, and the hidden ones remain hidden. The activity indicators of the visible neurons on the surface are rearranged as a result of an incoming external stimulus, whereas the activity indicators of the neurons hidden inside the mass are modified as a result of the cybernetics of the neural network model to achieve one of the equilibrium configurations. The controlled change also suffers from the uncontrollable fluctuation.

The situation described above looks very similar to the dynamics of a spin-glass model or the Ising model of matter subject to uncontrollable thermal fluctuations. The spin-glass model is a simplified model of matter of finite size in which the degrees of freedom of fundamental constituents, such as **atoms**, are approximated by dichotomous variables taking the value *+1* or *-1*. It is customary in physics to call such a dichotomous variable a spin variable, corresponding to the image of a top spinning right side up or upside down, so this simple model of matter is called a spin-glass model. The term "glass" refers to the fact that the spatial distribution of the fundamental constituents with spin cannot be aligned, as they are in the case of crystals. Neurons forming the neural network are certainly not aligned, and neither are the atoms forming a glass.

Now the process of imprinting the memory of an external stimulus might fall under the dynamics of a spin-glass model of matter of finite size in physics. The spin-glass model always suffers from the thermal fluctuations represented by the **Boltzmann distribution** with respect to a certain given temperature. We can fix the actual value of the temperature in such a way that the variance of the thermal fluctuation coincides with that of the fluctuation in the stochastic differential equation governing the increment of the generator potential. Suppose that you manually arrange the spin variables on the surface of the spin-glass so that you make up your own distribution pattern of spin variables *+1* or *-1*. This distribution pattern stands for an external stimulus. The given values of the spin variables on the surface and the actual values of spin variables inside the spin-glass are the initial values of all the spin variables. Those spin variables will suffer from controlled changes and settle into a certain local equilibrium configuration of spin variables *+1* and *-1* over the whole spin-glass. They will still undergo random changes due to the thermal fluctuation, though they remain unchanged in the mean. This local equilibrium configuration of spin variables *+1* and *-1* over the whole spin-glass given here is the imprinted short-term memory of the external stimulus represented by the given surface distribution pattern of the spin variables of the spin-glass.

We see that, in a sense, brain functioning is comparable to the thermodynamics of a spin-glass as long as the connectionist neural network model is used to describe the cybernetics of the brain. For that reason, we should give up the neural network model of cybernetics and look for another model of cybernetics, one that shows brain activity to be much more meaningful and colorful than the spin-glass.

Section 38
The Dendritic Network Model

We have proposed that brain functioning pertaining to consciousness and memory arises from cooperative physical phenomena taking place among the brain cells in the dendritic networks of the cerebral cortex. From the point of view of cybernetics, however, there is no significant difference between the neural network and the dendritic network, even though the latter is a much more refined and sophisticated system. Indeed, efforts have been made to describe the cybernetics of the dendritic network in terms of the stochastic differential equation governing the increment of some local dendritic membrane potential to show that the cybernetics of the dendritic network are analogous to that of a neural network. As long as the same principle of cybernetics is applied to both, we will inevitably see brain functioning as equivalent to the thermodynamics of a spin-glass, despite the obvious fact that the dendritic networks have a much finer structure. We need a new understanding of cybernetics in this context to develop a dendritic network model of brain functioning, just as one needs a new wine bottle for fresh wine.

Initially, a phenomenological principle was introduced to describe the cybernetics of the neural network because the mechanism controlling synapse connectivity among the neurons in the neural network was not known. The problem remains, however: we still need to know the real mechanism that controls the cybernetics of the dendritic network. To this end, it seems important to investigate the microscopic physical structure of the dendritic network. From the physical point of view, there exist in each location of the dendritic membrane biomolecules of high **electric dipole moment**—for example, lipoprotein molecules—whose ends are oppositely charged. Biomolecules, which constitute living matter, are **molecules** generally composed of many hydrogen atoms, oxygen atoms, carbon atoms, nitrogen atoms, sulfur atoms, phosphorus atoms, and so on. The positive poles of those biomolecules are aligned on the inner surface of the dendritic membrane, and they attract negatively charged atoms called negative ions. The negative poles are aligned on the outer surface, and they attract positive ions. The positively and negatively charged elements in the two separate interface regions immediately adjacent to the dendritic membrane manifest dynamics via the electromagnetic interaction. The regions where this interaction occurs are called Debye layers, bioplasma layers or perimembranous regions. Recall that plasma is the fourth phase of matter in which positively and negatively charged, fundamental constituents interact with each other and produce peculiar dynamics that resemble the cybernetics of living matter.

As shown in the preceding chapter, the dendritic network is a structure composed of the dendrites of several neurons entangled with each other in a highly sophisticated fashion by means of chemical synapses on the tips of the dendritic spines and by ephapses and tight junctions on the dendritic membranes. Those connecting nodes play the important role of microscopic gates through which different dendrites exchange neurotransmitters, ions, and electric potential. The neurotransmitters control the opening and the closing of the ion channels, and the ions flashing to and fro through their respective ion channels change the electric potential across the dendritic membrane. During this process, the most fundamental degree of freedom of the dendritic network is found to be the electric-potential difference distributed over the whole dendritic membrane of the dendritic network. Transmembrane polarization is the term for this fundamental degree of freedom. This transmembrane polarization affects the conformation of the membrane biomolecules, changing their effective charges and producing electric currents in the perimembranous regions immediately adjacent to the polarized dendritic membrane. In this sense, transmembrane polarization controls the electric current and, in turn, the distribution density of ions in the perimembranous regions; therefore, we call transmembrane polarization a control variable of the dendritic network to emphasize the cybernetics of the dendritic network. It determines the existence of a **field** over the whole dendritic network.

The distribution density of charged ionic elements in the perimembranous regions is governed by transmembrane polarization and undergoes controlled change, thus reflecting the cybernetics of the dendritic network. We call the distribution density of charged ionic elements in the perimembranous regions a state variable of the cybernetics of the dendritic network. It also determines the existence of another field over the whole dendritic network. We need to know how the activity of the dendritic network changes under the influence of the control variable.

Given the state variable as well as the control variable, we can investigate the cybernetics of the dendritic network. To make our cybernetics model of the dendritic network realistic, we take into account some established physical characteristics of the perimembranous regions immediately adjacent to the dendritic membrane. One characteristic is that the biomolecules of high electric dipole moment oscillate periodically (Fröhlich 1968). Even though the induced membrane-electric-potential oscillation has a small **amplitude**, it can be considered essential to the cooperative activity in the cybernetics of the dendritic network. Gray and Singer have recorded direct experimental evidence of such oscillations and discussed in detail their significance for distributed cooperative activity of the dendritic network in the primary visual cortex (Gray and Singer 1989). This small but periodic oscillation provides the key concept in the creation of a cybernetics model of the cooperative activity in the dendritic network.

First, let us consider an extremely idealized case in which the oscillations of the membrane-electric-potential in every location of the dendritic network are perfectly synchronized. The control variable manifests a synchronized harmonic oscillation with a common **frequency** everywhere. In this extremely idealistic case of synchronized harmonic oscillation, the distribution density of charged elements in the perimembranous regions immediately adjacent to the dendritic membrane remains uniform, and there are no ionic currents in the dendritic network.

Let us proceed next to a less idealistic case in which the oscillation of the membrane-electric-potential in each location of the dendritic network manifests a deviation from the synchronized harmonic oscillation. The deviation might not be uniform, and the control variable shows oscillations with different **phases** in different locations. The control variable illustrates a phase wave spreading over the whole dendritic network, which may be likened to **de Broglie**'s **matter wave** in quantum theory. Recall that the matter wave, that is, a phase wave of the **matter field**, controls the dynamics of a **quantum**. In this cybernetics model of the dendritic network, a phase wave of the dendritic membrane-electric-potential field controls the dynamics of charged elements in the perimembranous regions. This formal coincidence between the cybernetics of the dendritic network and the dynamics of the matter field might seem accidental, but the following chapter will reveal that the present cybernetics model of the dendritic network can be obtained in terms of the collective mode and the macroscopic wavefunction of quantum field theory applied to the fundamental processes of the brain. *Brain activity is essentially of a quantum physical nature.*

Think of the controlled change of the distribution density of charged elements in the perimembranous regions as a flow or a current in those regions described as a field of **velocity vectors** of charged elements over the dendritic network. This field of the velocity vectors over the dendritic network, called a density current, gives us the electric current in the perimembranous regions if it is multiplied by the distribution density of charged elements.

[Distribution Density][Density Current]=[Electric Current]

In symbolic notation, we have

$$\rho \mathbf{v} = \mathbf{J}$$

where ρ stands for distribution density, $\mathbf{v}$ for density current, and $\mathbf{J}$ for electric current. Thus, we see immediately that the state variable of the dendritic network, that is, the distribution density of charged elements in the perimembranous regions, is governed by the equation of continuity:

{Time Derivative of}[Distribution Density]

=-*{Divergence of}[Electric Current]*

Or symbolically,

$$\frac{\partial \rho}{\partial t} = -\nabla \cdot \mathbf{J}$$

The equation means this: *The rate of temporal change of the distribution density of charged elements balances with the rate of the spatial change of charged elements moved along by the electric current.*

Recall the most idealistic case of the cybernetics of the dendritic network, presented earlier. In such a case, the control variable manifests a synchronized harmonic oscillation with a common frequency everywhere, and the state variable takes a common value everywhere, so no electric current is generated in the perimembranous regions immediately adjacent to the dendritic membrane. In a corresponding manner, a vanishing electric current causes a vanishing density current. Of course, this might be the result of a control mechanism inherent in the cybernetics of the dendritic network. The control variable undergoes an oscillation everywhere, and it rapidly changes even though it is synchronized and tuned everywhere. To find a model control mechanism, therefore, we have to look for some quantity that vanishes in such synchronized oscillations and might be directly connected to the vanishing density current. Among the many possibilities, the most plausible might be the phase of the control variable manifesting oscillations in every location. In the case of synchronized oscillations, the phase of the control variable also takes a common value everywhere, just as the state variable does. There, the vanishing quantity in question might be identified as the gradient of the phase of the control variable, that is, a field of **vectors** pointing in the direction of the steepest descent of the phase change in each location of the dendritic network.

In the less idealistic case, in which the control variable acts as a phase wave spread over the whole dendritic network, the density current might not vanish, but this control mechanism would still work. In this simple model control mechanism of the cybernetics of the dendritic network, we would find that the density current of the charged elements in the perimembranous regions is controlled in each location so that it is proportional to the gradient of the phase of the control variable.

[Density Current] ∝ *{Space Derivative of}[Phase]*

In symbolic notation, this becomes

$$\mathbf{v} \propto \nabla\Theta$$

where Θ stands for phase of the control variable. The analogy to quantum theory becomes more explicit now, and the equation of continuity yields one of the fundamental equations describing the cybernetics of the dendritic network:

{Time Derivative of}[Distribution Density]
=-[Constant]{Divergence of}
([Distribution Density]{Space Derivative of}[Phase])

Or symbolically,

$$\frac{\partial \rho}{\partial t} = -\ell\nabla\cdot(\rho\nabla\Theta)$$

The density current is always equal to the gradient of the phase of the control variable times the constant ℓ. This constant of proportionality is called, more specifically, the Wiener constant. It seems worthwhile to note here that the control mechanism given by this equation has been observed experimentally in the cybernetics of amoeba-like monocellular organisms (Yasue 1988).

Section 39
Wavefunction/Order Parameter

The equation we finally obtained in the preceding section describes how the state and control variables are coupled in the present cybernetics model of the dendritic network. However, to adequately illustrate the model control mechanism of the cybernetics of the dendritic network, another fundamental equation is needed. Since the distribution density of charged elements in the perimembranous regions is the state variable, we may suppose that it obeys certain dynamics subject to specific physical laws in both electromagnetism and hydrodynamics. Those physical laws can all be derived from a least action principle, that is, a principle claiming that the dynamics of natural phenomena are optimized so that the physical quantity called the action is minimal. The least action principle is understood to be the most fundamental form of the natural law in physics.

Let us forget for the moment the existence of the control variable and consider the dynamics of the distribution density of charged elements in the perimembranous regions. The least action principle claims that the dynamics is governed by the hydrodynamic equation:

[Distribution Density]{Temporal Change of}[Density Current]
=[External Force]-[Pressure]

In symbolic notation,

$$\rho \frac{\partial \mathbf{v}}{\partial t} = \mathbf{F}_{ex} - \mathbf{P}$$

where $\mathbf{F}_{ex}$ stands for external force and $\mathbf{P}$ for pressure. The above equation means this: *The temporal change in the density current in each location of the perimembranous regions times the distribution density of charged elements is equal to the external force minus the pressure.*

In general, two factors contribute to the temporal change of the density current: one, its time derivative and, two, the difference between the incoming and the outgoing density current as represented by a directional derivative of the density current along its own flow direction. The pressure is given by minus the gradient of some quantity representing the congestion of the charged elements in each location of the perimembranous regions. The population potential, as it is called, depends on the distribution density of the charged elements, that is, the state variable. The external force is also given by minus the gradient of some potential energy representing the energy gain due to the external action. That will be called the external potential.

The density current is governed by the control variable in such a way that it equals the gradient of the phase of the control variable multiplied by the Wiener constant; therefore, the hydrodynamic equation introduced above can be rewritten in terms of the phase of the control variable, which will be called simply the **phase**:

[Wiener Constant]{Time Derivative of}[Phase]
= -([Wiener Constant]^2/2)({Space Derivative of}[Phase])^2
+[Population Potential]-[External Potential]

Or symbolically,

$$\ell \frac{\partial \Theta}{\partial t} = -\frac{\ell^2}{2}|\nabla \Theta|^2 + U - V_{ex}$$

where U and V_{ex} denote population and external potentials, respectively.

The equation is explained as follows: *The time derivative of the phase multiplied by the Wiener constant is equal to minus one half of the squared gradient of the phase multiplied by the Wiener constant squared plus the population potential minus the external potential.*

We regard the preceding equation as another fundamental equation of the cybernetics governing the control variable. To our astonishment, this equation is of the same form as **Schrödinger**'s wave equation in Bohm's form:

- The angle of the hand of each Erwin meter corresponds to the phase.

$$\theta \to \Theta$$

- The **Planck constant** corresponds to the Wiener constant.

$$\hbar \to \ell$$

- The **mass** corresponds to unity.

$$m \to 1$$

- The potential energy corresponds to the external potential.

$$V \to V_{ex}$$

- The quantum potential corresponds to the population potential.

$$Q \to U$$

It should be explained here that population potential is global in the sense that quantum potential is global, though both potentials are local in their functional forms. The essential point is this: Both population potential and quantum potential are given by the distribution density and the probability density, respectively, in terms of local mathematical manipulations, such as differentiation. In this sense, they are local functions of densities; however, neither the distribution density nor the probability density can be determined locally. We need global information to build up those densities. Indeed, those densities are given by integrating the initial values over the whole space; this mathematical manipulation is totally global.

The first fundamental equation of the cybernetics model of the dendritic network was the equation of continuity. In quantum theory, the equation of continuity played an important role in casting Schrödinger's wave equation into Bohm's form. Here, we find again a perfect parallel between dynamics in quantum theory and cybernetics in control theory. The two fundamental equations of cybernetics—that is, the equation of continuity for the state variable and the hydrodynamic equation for the control variable—can indeed be brought together in a single and simple equation for a complex-valued quantity corresponding to the Schrödinger equation for the wavefunction. This complex-valued quantity, given by identifying the square root of the distribution density and the phase with its absolute value and argument, respectively, will be called the wavefunction or the order parameter of the cybernetics of the dendritic network.

Finally, in the same manner that we obtained the Schrödinger equation from Schrödinger's wave equation in Bohm's form in quantum mechanics, we obtain the fundamental equation for the order parameter that governs the cybernetics of the dendritic network under the optimal control principle.

The cybernetics of the dendritic network is described by the order parameter in such a way that the following Schrödinger-like equation holds:

[Imaginary Unit][Wiener Constant]
{Time Derivative of}[Order Parameter]
=-([Wiener Constant]^2/2)
{Divergence of}{Space Derivative of}[Order Parameter]
+[External Potential][Order Parameter]

In symbolic notation, we have

$$i\ell\frac{\partial\Psi}{\partial t} = -\frac{\ell^2}{2}\nabla\cdot\nabla\Psi + V_{ex}\Psi$$

where Ψ denotes order parameter.

Once the evolution of the order parameter is found by solving this equation for a given external potential and a given initial value, the controlled changes in the degrees of freedom of the dendritic network, that is, the cybernetics, can be illustrated immediately as follows: The distribution density of charged elements, that is, the state variable, is given by the **absolute square** of the order parameter; the phase of the oscillating membrane electric potential, that is, the control variable, is given by the phase factor (i.e., argument) of the order parameter.

Thus, in the present optimal control model, the cybernetics of the dendritic network may be described by the above fundamental equation, though it has been derived phenomenologically from some of the physical facts known about the perimembranous regions immediately adjacent to the dendritic membrane. It seems worthwhile to note here that Gordon Globus has developed the post-modern theory of the brain in which the present cybernetics model is employed as a *wave formulation.* This emphasizes his interesting philosophical image of the human brain as spontaneous, eruptive, and symmetry-increasing, in contrast to the mechanistic image that portrays the brain as computer-like, clock-driven, step-wise, and logical (Globus 1995). Globus proposes that the brain does not exhibit information processing; rather it produces influence that fans out to the whole. It is not programmed but self-tuning, flowing, fluctuating, convolving, interpenetrating, evolving toward least asymmetry. The cybernetics model of the dendritic network, which we have developed so far, indeed may manifest those peculiar aspects of brain functioning.

Section 40
Wave Cybernetics of the Brain

Extracting the essence of the cybernetics model of the dendritic network developed in the preceding section, we arrive at the following scheme of optimal control strategy realized in the cybernetics of brain activity:

1. The oscillating control variable is tuned over the whole system and contains the global symmetry of the **implicate order** enfolded to the system. The distribution of the tuned oscillations over the whole system manifests a phase wave, which controls or guides each state variable.

2. Each state variable evolves due to local interaction with the globally tuned phase wave. The resulting time evolution of the state variable exhibits nonlocality arising from the global symmetry of the implicate order of the system contained in the control variable.

3. The cybernetics of the system with state and control variables is described by the order parameter or the wavefunction subject to a Schrödinger-like wave equation. The time evolution of the state variable is obtained as the absolute square of the order parameter, and that of the control variable as the argument of the order parameter.

The above scheme gives rise to nonlinear and nonlocal cooperative cybernetics, even though its fundamental equation is apparently linear and local due to its Schrödinger-like form. The apparent linearity and locality of the fundamental equation make the analysis much simpler, while the essential nonlinearity and nonlocality underlying the whole framework adequately capture the cybernetics of the dendritic network. In other words, both the nonlinearity and the nonlocality essential to the cybernetics of brain activity are well reflected in this scheme of optimal control strategy, even though the fundamental equation governing the cybernetics is a linear and local equation that is easy to handle. This might be the nonlocality Pribram sought in the cooperative physical phenomena taking place in the dendritic network (Pribram 1991). For a different form of population potential, however, the essential nonlinearity of the cybernetics of brain activity becomes apparent even in the Schrödinger-like fundamental equation. The Schrödinger-like wave equation turns out to be a nonlinear wave equation such as the Ginzburg-Landau equation or the Schrödinger-Langevin equation (Yasue 1978). Those nonlinear versions of the Schrödinger-like wave equation may permit both laminar (regular) and turbulent (chaotic) flow of the order parameter. The essential and apparent nonlinearity of the cybernetics of brain activity manifests unpredictability due to the phenomenon of chaos (Globus 1995).

As the phase wave of certain oscillating degrees of freedom controls the cybernetics, the above scheme will be called *wave cybernetics*. It may be understood in light of **Bohm**'s natural philosophy, which maintains that the universe, that is, the whole system, is filled with a tuned **wave**, from which each element in the universe is determined in its evolution (Bohm 1980).

Let us see the way in which the dendritic network is influenced by an external potential (e.g., a sensory input) given by the activity of a neural network in one of its equilibrium configurations. Consider a dendritic network containing a few neurons with axons belonging to a neural network. Those neurons mediate between the neural network and the dendritic network. The electric-potential distribution on the dendritic membrane of those neurons is determined by their activities as elements of the neural network, and this electric-potential distribution may be regarded as external potential (sensory input) in the cybernetics of the dendritic network, that is, the controlled change of charged elements in the perimembranous regions immediately adjacent to the dendritic membrane. Namely, the external potential appearing in the Schrödinger-like wave equation for the wave cybernetics of the dendritic network participates in the transmission of the local activity of the neural network to the dendritic network. The external potential is a control factor in the cybernetics in that it prompts changes in the dendritic network. Since the cybernetics of the dendritic network has its own characteristic time scale, which is much shorter than that of the neural network, any local activity of the neural network can be regarded as time-independent when seen as an external potential. The characteristic time scale of the cybernetics of the dendritic network is estimated to be the period of fundamental electric-potential oscillations of certain biomolecules with high electric dipole moment, on the order of several picoseconds. Any electric-potential fluctuation on the dendritic membrane with a characteristic time scale much longer than those values can be regarded as the external potential controlling the cybernetics of the dendritic network.

Any local activity of the neural network whose activity distribution is controlled so as to minimize the total configuration energy with respect to a given external stimulus imposed on the visible neurons in the neural network will determine the electric-potential distribution on the neurons mediating between the neural network and the dendritic network, that is, the external potential affecting the cybernetics of the dendritic network, which manifests immediate response to the external potential with a characteristic time scale of several picoseconds. This response of the dendritic network due to the cybernetics realized in the perimembranous regions is described by the Schrödinger-like wave equation, so it exhibits the nonlocality typical of wave cybernetics. The output of this nonlocal response of the dendritic network can be understood as an influence of the cybernetics of the dendritic network on the local activity of the neural network. This influence is produced by the stationary distribution pattern of the charged elements in the perimembranous regions immediately adjacent to the dendritic membranes of several neurons that belong to both the dendritic network and the neural network. Thus, the nonlocal, cooperative phenomenon of wave cybernetics taking place in the dendritic network influences or modulates the macroscopic cybernetics of the neural network. Highly advanced functions of the brain, such as consciousness and its qualitative activities, may be due to the influence of the wave cybernetics of the dendritic network on the macroscopic information processing realized by the neural network.

Section 41
Holonomic Brain Theory

It is possible that memory and consciousness arise from certain microscopic physical phenomena in the cryptic cosmos of the brain. We are going to dive into the microscopic world inside the cell membrane in search of the real nature of memory and consciousness. It is a world far beyond the scope of our imaginations, so different is it from our daily experience. It is a world governed by the most fundamental laws of physics, that is, the laws described by quantum theory. That is the reason we imparted a minimum knowledge of quantum theory in Chapters 1, 2 and 3 of this volume. We will take leave of modern cerebral physiology, brain science, and molecular biology as we commence our journey. No vehicle other than quantum theory can bring us to the terra incognita of the microscopic world of the brain. Giving us a hearty send-off are several brain scientists, including John Eccles, Gordon Globus, Stuart Hameroff, and Karl Pribram, who has advocated the nonlocal theory of memory since Lashley passed away. Why is Pribram seeing us part of the way? There is a reason, of course.

As we have seen, Pribram came to believe firmly in the nonlocality of memory in the 1950s while investigating the grandfather cells of the visual area. He speculated that the microscopic physical phenomena in the dendritic network might be fundamental processes in the production of memory and consciousness. An accomplishment in the field of physics provided support for his work. The boon came from the successful reproduction of a stereoscopic image by means of a laser beam. This stereoscopic image is called an optical holograph. It had already been proposed theoretically by physicist Dennis Gabor in 1947, but its realization was contingent upon the emergence of laser light fifteen years later. Prior to that, it was difficult to obtain a source of light with a homogeneous phase.

A laser is a device that can emit a specific form of light in which numerous photons with a definite **wave number** display the same wave motion as that of the **electromagnetic field**. This is accomplished by a collective motion of the **electrons**, called the collective mode, in a large number of atoms interacting with the electromagnetic field. Recall that a global harmony or order emerges in the collective mode of the creation and annihilation dynamics of energy **quanta** of the matter field and the electromagnetic field, that is, electrons and photons. Due to the presence of this global order, the creation and annihilation dynamics of electrons and photons in the laser device is so elaborately tuned that an extremely large number of photons can be created in a single wave motion of the electromagnetic field with a definite **wavelength** and phase. Then, ideal monochromatic light made of only specific wave motions of the electromagnetic field with a single and definite wavelength and phase can be produced by a laser. This is a laser beam.

The principle behind the optical holograph, that is, the stereoscopic image, is neither too sophisticated nor difficult to understand. It uses the **interference** effect of waves, as in the double-slit experiment for electrons or photons, which we discussed in Chapters 1 and 2. The interference of the wave motion of the electromagnetic field is used in the case of optical holography. We know that the real nature of light must be grasped from the point of view of quantum theory. A laser beam is nothing but specific light corresponding to a great many photons created in a single wave motion of the electromagnetic field with a definite wavelength and phase. Therefore, it can be regarded as a high intensity, ideal monochromatic light beam with a common wavelength and phase when considered as a wave. Such ideal monochromatic light is said to be coherent light.

Suppose a laser beam emitted from a laser device is split into two beams by a beam-splitter, usually made of a half-silvered mirror, which is a mirror partially coated with silver to make it half-transparent and half-reflective. One of the split beams irradiates an object, and the light that scatters from the object is guided to and recorded on a film. The other split beam is guided directly to the same film and recorded simultaneously with the scattered light. The spatial difference be tween the point on the surface of the object and the point on the film is measured in the phase difference between the two beams. Those phase differences are recorded as a striped interference pattern due to the interference effect between the scattered light and the direct beam, which plays the role of a reference wave. The striped pattern on the film, produced by the interference effect of light, is called a *hologram*, and it contains stereographic information about the object. For an ordinary photograph, only the intensity distribution of light scattered from the object is recorded on the film, and the image reproduced by passing a reference light through the recorded film remains flat. In optical holography, however, a stereoscopic optical image can be reproduced by passing a reference laser beam through the hologram.

Although a hologram on the film itself is nothing like the object it represents, the information about the phase differences between the two beams is stored in the whole interference image of the hologram. This stored order is Bohm's "implicate" or "enfolded" order. Even a small portion of the hologram contains essentially all the information concerning the phase differences between the beam scattered from all points on the surface of the object and the reference beam. In other words, the stereographic optical image suffers no defect, even if a part of the hologram on the film is lost. (Of course, some reduction in resolution occurs.) Thus, *the whole is in the part.* This astonishing feature of optical holography was noted by Karl Pribram, who had been searching for a yet unknown mechanism of memory that might account for nonlocality in the brain. That was in the year 1965.

On the one hand, there was the physical phenomenon of optical holography obtained first by applying quantum field theory to the electromagnetic field. On the other hand was the nonlocal mechanism of memory in the cerebral cortex. What kind of connection might exist between them?

To see this, we might liken the hologram to the memory trace, that is, the *engram* of an object. The hologram as a memory of the solid shape of an object reproduces an optical stereoscopic image without defect even if part of the hologram is lost. However, the larger the loss, the vaguer and dimmer becomes the reproduced optical image as the reference laser beam passes through it. It is perhaps not too much to say that the phenomenon discovered by Lashley and Pribram in the experiments they performed on the brain to find the trace of memory, is, indeed, characteristic of the hologram. The phenomenon was nonlocality. That is, there appeared to be no specific loss of memory or function when part of the cerebral cortex was destroyed, but the greater the destruction, the more globally degraded the memory or function became. No indication of the alleged localization of either memory or brain functions was found; damaging specific cortical pieces did not annihilate the memories or the functions supposedly associated with them. However, if damage to the cerebral cortex becomes more widespread, disorder appears in overall brain functioning.

Can you imagine the look on Pribram's face when, *by chance*, he found an article reporting the success of optical holography? It is not difficult to see an analogy between the hologram on a film and memory in the cerebral cortex. In light of the results of his experiments pertaining to brain lesions, Pribram intuited that memory is similar to a hologram. The outcome of the work he had started with Lashley was no longer a mystery: *Memory is stored just like a hologram.* Of course, it was evident to Pribram at that time that there existed no optical holographic device in the brain, nor did the cerebral cortex become a film on which to record a hologram. Pribram's insight, however, was that there exists a certain yet unknown physical wave in the cerebral cortex, that a hologram is produced by this physical wave, and that this hologram stores memory in the brain. He called this new holographic mechanism a *neural holography* and anticipated that it might be the nonlocal mechanism of memory.

Contrary to his belief, many people claimed that his idea of neural holography was unrealistic and, without correctly understanding it, interpreted it as a literal instantiation of optical holography. Most of the brain scientists at that time could not believe that such optical phenomena were fundamental to brain activity, nor could they imagine a holographic mechanism other than the optical one. That was in the 1960s, when holographic devices, such as those used in electron wave holography, were still new, and ultrasonic holography had not yet been invented. (In electron wave holography and ultrasonic holography, the wavefunction of the electron and the ultrasonic wave are used, respectively, in place of coherent light.) The scientific community at that time did not believe in the existence of a certain wave phenomenon in brain tissue that gives rise to a holographic mechanism. Furthermore, Pribram, who had started out as an excellent brain surgeon, was still a brain physiologist devoting all his time to experimental research. He could not immediately develop a theory of neural holography in a form convincing to others, even if by intuition he understood memory as a hologram. Many people said that Pribram's neural holography was not well substantiated and that it was a childish idea. His situation was similar to de Broglie's when he proposed his profound concept of the matter wave.

Alone and against great resistance, but always full of spirit, Pribram strove to substantiate his theory. Understanding that no "covering fire" could be expected, he proceeded on his own to develop a basic theoretical framework of neural holography. It was called *holonomic brain theory* (Pribram 1991). The concept of the dendritic network played a key role. We have already explored the cybernetics of the dendritic network, as noted by Pribram. Recall that we studied the fundamental oscillation of the electric potential of the dendritic membrane and discovered that the dendritic network manifests essentially nonlocal cybernetics called wave cybernetics, as described by the Schrödinger-like wave equation for the order parameter. Although the nonlocal cybernetics of the dendritic network was obtained phenomenologically from the microscopic, material structure of the dendritic network, its formulation resembles that of quantum mechanics in terms of Schrödinger's wave equation. Therefore, the same holographic mechanism might apply to both the cybernetics of the dendritic network and the dynamics of electron wave holography. That is, the wave cybernetics of the dendritic network might provide a holographic mechanism derived from the interference effect applied to a Schrödinger-like wave equation in the wave cybernetics of control theory, and this same holographic mechanism might appear in the dynamics of electron wave holography based on the Schrödinger equation in the wave mechanics of quantum theory. This holographic mechanism might be the basis of neural holography.

Thus, a candidate for neural holography has been found within the realm of holonomic brain theory in terms of the wave cybernetics of the dendritic network. It results from the interference effect among wave phenomena taking place in the perimembranous regions immediately adjacent to the dendritic membrane, but, as expected, it is different from optical holography. Perhaps no one believed that optical holography could be realized literally in brain tissue. Naturally Pribram looked for other possibilities in the microscopic physical phenomena occurring in the dendritic network (Pribram 1991). This question arises: *Is there really no room in the brain for optical holography?* In the next chapter, we shall investigate a physical theory that might help us to understand memory and consciousness as natural phenomena relevant to the first principles of quantum mechanics and quantum field theory. *Astonishing though it may be, we find there a firm possibility of quantum theoretical optical holography in terms of coherent photon emission from a collective mode of water molecules in and around microtubules in the cytoskeletal structure and extracellular matrices* (Jibu et al. 1994).

Part Three

A Quantum Approach to Brain Functioning, Consciousness and Life

Chapter 7
Quantum Theory of the Brain

Section 42
Hiroomi Umezawa

At the same time that the success of optical holography inspired Pribram's concept of neural holography, a Japanese physicist speculated that the fundamental process of **memory** in the brain could not be understood correctly without incorporating quantum mechanical concepts, especially, **quantum field theory**. That physicist was Hiroomi Umezawa, successor to Hideki Yukawa who had initiated a quantum field theoretical investigation of the fundamental constituents of **matter**. Umezawa had trained his own successors in Italy, the United States and Canada. His open-mindedness and enterprising spirit distinguished him from other Japanese physicists of his day. Umezawa pursued a wide range of interests, not only in physics but in other human activities as well. Indeed, Umezawa often played the violin while studying physics. One essential question occupied his mind: *What is the mechanism of memory that underlies consciousness and gives rise to all the humanities, including philosophy and the arts?*

What kind of mystery is insinuated in our attraction to the mechanism of memory? What silhouette of quantum field theory did Umezawa see in memory? Let us play for a while with Umezawa within this mysterious realm of memory. Take a coffee break and remember anything that happened while you were in high school. No doubt, you remember something clearly. Recall a memory. You have spent most of your time since then engaged in current, day-to-day events, without recalling incidents from your youth. Your consciousness has been driven by kaleidoscopic changes, and you might not remember previous events in your life if you were not asked here. Whether or not those memories have been recalled, they have never been annihilated, as you can see.

Memory, therefore, appears to have an extremely stable existence. As long as the trace of memory, that is, the *engram*, exists in the brain as a certain physical entity, its stability should be understandable within the realm of physics. In the preceding chapter, we investigated the possibility that the mechanism of memory arises from certain cooperative physical phenomena on a microscopic scale in the cerebral cortex. The stability of memory becomes an important factor to consider when attempting to identify the yet unknown physical processes taking place there. However, as far as physics is concerned, only a restricted number of physical entities are extremely stable. The reason is that any physical process that loses **energy** cannot be maintained. Of course, if all the actions from the external world are shut off and the physical system in question becomes isolated, then energy is conserved and the physical processes therein can be maintained. However, the isolated brain, not interacting with the external world, can no longer be said to be a brain. Memory can remain stable even under circumstances that allow external stimuli to impinge upon it.

Thus taking only the stability of memory into consideration, we see a limitation on the number of physical phenomena that might account for the fundamental process of memory. As a matter of fact, a physical process that remains stable without losing energy even under such a non-isolated (i.e., open) circumstance emerges first in quantum field theory. Namely, it is the dynamics of a sub-microscopic **field** entity in the energy eigenstate with the minimum energy eigenvalue; this is known as the vacuum state. As was found in quantum field theory, only a restricted number of **wave** motions of a field are realized on a sub-microscopic scale, and those correspond to energy eigenstates. Therefore, the wave motion of the field entity in the vacuum state with the minimum energy eigenvalue does not lose energy. In other words, *the vacuum state represents a perfectly stable physical entity in quantum field theory.* From this fact we can surmise that Umezawa got the idea "memory is a vacuum state" when he came to believe that quantum field theory was indispensable in investigating the fundamental process of memory in the cerebral cortex.

A problem is evident here: If memory is nothing but a vacuum state of a certain sub-microscopic field entity in the brain, it might turn out to be a unique vacuum state, as in the case of the **electromagnetic field**; that is, there is only a single vacuum state with the minimum energy eigenvalue containing no **photons** at all. Consequently, only a single memory could be stored. However, actual memory seems much more rich and diverse in both quality and quantity. Furthermore, an actual memory might be extremely stable but not perfectly so, and it can be annihilated as time passes. If memory is given by the vacuum state, then it would never vanish; that is not characteristic of memory. Was Umezawa's excellent idea no more than a metaphor illustrating one of the mysterious characteristics of memory, that is, its stability? There is no need to worry about that. Picture the smiling face of Umezawa and hear him saying, "*The real pleasure of quantum field theory is the interaction, my friends.*" He sought interacting fields to account for memory.

Investigating the wave motion of the electromagnetic field within the realm of quantum field theory, we find that it can be regarded as an energy quantum called a photon. The electromagnetic field is not subject to any action from other **degrees of freedom**. According to **quantum theory**, without interaction in such a case, the vacuum state of the field entity would be unique and perfectly stable. However, an isolated field entity is rather unusual in quantum field theory; in most cases a field entity suffers from external action, and its own self-action manifests much more sophisticated dynamics than the dynamics of an isolated field without self-action. When one tries to apply quantum field theory to such an interacting field, the vacuum state cannot be unique. In general, an infinite number of vacuum states emerge. Furthermore, each of those infinitely many vacuum states comes to be merely a false vacuum state that approximates a minimum energy eigenstate due to the interaction, so instead of remaining perfectly stable, each one changes into another as time passes. If the fundamental process of memory in the brain is described by a certain wave phenomenon in quantum field theory, the field entity taking part therein naturally suffers from interactions with other physical degrees of freedom. Umezawa, reflecting on the existence of such interactions, felt in his bones that memory would be a vacuum state. Once such a vacuum state of a certain field entity in the sub-microscopic world of the brain is regarded as memory, then both the infinite variety and the imperfect stability of memory might be explained within the realm of quantum field theory. Furthermore, it can be concluded also that memory manifests **nonlocality** because the vacuum state of a certain field entity is regarded as a stable physical object with macroscopic spatial extension. Recall that the collective mode—in which the creation and annihilation dynamics of energy **quanta** all fall into the vacuum state—manifests a nonlocal order on the macroscopic scale.

Section 43
The Quantum Field Theory of the Brain

At the same time that Karl Pribram in the United States conceived the idea that memory might be a hologram manifesting stability and nonlocality, Umezawa in Italy intuited that memory was a quantum vacuum state also manifesting stability and nonlocality. It is not an uncommon occurrence in science that people working independently arrive at a similar conclusion, by "coincidence," as did Pribram and Umezawa. The more original and epoch-making the breakthrough, the more frequently such a "meaningful coincidence" seems to occur. For example, at the beginning of this century, Jean Perrin in Paris got the idea that an **atom** would be a sub-microscopic solar system in which negatively charged **electrons** go around a central core made of positively charged, heavy

particles, and Hantarou Nagaoka in Tokyo proposed independently a mechanical model of the atom in which an atom was regarded as a sub-microscopic solar system. (**Niels Bohr** presented a new idea on the structure of the atom based on Nagaoka's solar-system model of the atom and the quantum condition.) **Carl Gustav Jung** was a famous psychiatrist in the suburbs of Zürich who shed light on the phenomenon of "meaningful coincidence." He called it **synchronicity** and attributed it to the **collective unconscious** underlying all human consciousness. Late in life, he collaborated with **Wolfgang Pauli** to substantiate this yet unknown entity called the collective unconscious. Pauli was an outstanding physicist and Nobel laureate in Switzerland who, together with Werner Heisenberg in Germany, first proposed quantum field theory. It has been said that Pauli explored the electromagnetic field as a physical entity that might account for the mechanism of Jung's collective unconscious and synchronicity. He looked for theoretical support in terms of quantum field theory. It is unfortunate that we do not know what kind of quantum field theory of the collective unconscious Pauli might have proposed because the results of his collaboration with Jung were concealed after Pauli passed away.

Pauli and Jung sought the mechanism of the collective unconscious that connected a number of human brains across time and space. The intention of their research was not to grasp the fundamental mechanism of consciousness and memory in a single brain. One might say that Umezawa initiated scientific investigation into the mechanism of consciousness and memory in the individual brain within the physical framework of quantum field theory. Between 1967 and 1979, Umezawa published three papers pertaining to the quantum field theory of brain. The first paper was written in collaboration with one of his Italian colleagues, Luigi Ricciardi, and published in a technical journal on cybernetics (Ricciardi and Umezawa 1967). The second and third papers were published in 1978 and 1979 in technical journals of theoretical biology and fundamental physics, respectively, after Umezawa had moved to Edmonton, Canada (Stuart et al. 1978; 1979).

In the 1960s and 1970s, the majority of brain scientists supported the neuron doctrine. They looked for the fundamental process of brain activity in the neuron, which, from the point of view of brain physiology, was considered a materialistic unit of the brain. It was customary in brain science, physical chemistry, and **molecular biology** to concentrate on material constituents that manifest continual existence or appearance. It was consistent with the tradition, therefore, to seek a mechanism of consciousness and memory in the material constituents of the brain. Umezawa was different, however. Since his youth, he had been attracted to philosophy, physics, mathematics, and psychology, and, wishing to have a clear view of the essential aspects of the human brain, he applied an interdisciplinary approach to the study of the human brain. He became a world-famous specialist in quantum field theory and elementary particle physics and established a school in Naples, Italy. There he was able to investigate fundamental processes in the sub-microscopic world of the brain from a drastically different point of view, one not prejudiced by the neuron doctrine (Ricciardi and Umezawa 1967).

In quantum field theory, one must take into account not only the fundamental constituents of matter but also the physical processes that arise from the sophisticated interactions among those constituents. The physical processes actually came to be regarded as quasi-fundamental constituents of matter due to the basic principles of quantum theory; that is, a large composite system, consisting of many fundamental constituents of matter, exhibits a behavior at the macroscopic level that is totally different from properties of the individual constituents (Umezawa 1993). This is essentially due to the action of the ordered dynamics of collective modes emerging from interactions among the individual components. Once created, such ordered physical processes constitute a phenomenologically distinct mode, thus enabling us to treat them as distinct physical entities. In this sense, they are quasi-fundamental constituents of matter.

The basic principles of quantum field theory provided the first account of the physical processes of the wave motion of fields as certain particle-like, material constituents. Recall the mesons predicted by Yukawa (1935) as elementary particles arising from the wave motion of the force field between nucleons (i.e., **protons** and **neutrons**). In Umezawa's mind, the neurons and the various biomolecules that constitute neurons were no more than redundant degrees of freedom playing only subsidiary roles, just as nucleons do in the case of mesons. Umezawa regarded the brain as a large composite system, consisting of many fundamental constituents of matter, and he suggested that the new quasi-materials, created as collective modes of the creation and annihilation dynamics of this large composite system with sophisticated interactions, become essential in brain functioning. Recall the performance of a synchronized swimming team: What the audience sees and what makes a skillful team are not the movements of each swimmer but a harmony or order in the total movements of and interactions among all the swimmers on the team. Umezawa thought that it is not the dynamics of the individual constituents but the ordered dynamics of collective modes of the large composite system of the brain, governed by the basic principles of quantum field theory, that plays an important role in realizing the highly advanced brain functions. For Umezawa, the real pleasure of quantum field theory was exploring the interactions among various fields.

In 1978 and 1979, Umezawa published his second and third papers in technical journals pertaining to theoretical biology and fundamental physics (Stuart et al. 1978; 1979). This time he collaborated with two physicists from Ireland and Japan: Iain Stuart and Yasushi Takahashi. Stuart had been investigating the foundation of the quantum theory of measurement in relation to human consciousness. Takahashi had achieved a world-wide reputation as an expert in quantum field theory after having discovered the famous Ward-Takahashi identity, which describes important aspects of the dynamics and the invariance of interacting fields. More than ten years had already passed since Umezawa got the inspiration that the mechanism of memory and consciousness in the cerebral cortex might be given by certain physical phenomena understood only by quantum field theory. By then brain scientists had offered some experimental evidence and theoretical considerations that supported his basic idea about memory as vacuum states. A type of nonlocal brain activity known today as synchrony or synchronous events had been discussed (John 1972; John et al. 1973; Bartlett and John 1973; Sayers et al. 1974; Sayers and Beagley 1974). Hence, in the third paper published in 1979, Umezawa, Stuart, and Takahashi examined and analyzed the experimental and theoretical evidence. That paper described the fundamental process of memory in the brain within the framework of quantum field theory. They focused primarily on experimental examinations and theoretical considerations that supported the nonlocality of memory. Pribram's nonlocal doctrine of memory, which incorporated quantum concepts such as holography, was highly regarded by the three authors.

Figure 10. Hiroomi Umezawa, Yasushi Takahashi (left) and Iain Stuart (right).

If memory is stored as a vacuum state of certain fields in the sub-microscopic world of the brain, as Umezawa envisioned, quantum field theory tells us that *the wave motion of fields in such a vacuum state manifests strong correlation and order extending to the macroscopic scale.* In other words, memory manifests nonlocality, and the trace of memory, that is, the *engram*, is nothing else but the wave motion of certain yet unknown fields in the vacuum state.

Was there a meaningful coincidence between Pribram's insight into the holographic character of memory and Umezawa's intuitive grasp of memory as a vacuum state? Might this be an example of synchronicity arising from the collective unconscious, the phenomenon suggested by Jung? It is clear at least that the deep thoughts of both Pribram and Umezawa offer a new glimpse of nature.

Section 44
The Sub-Microscopic World of Cytoplasm

Let us proceed to the terra incognita of the sub-microscopic world of the brain in which Umezawa's quantum field theory of the brain provides a deeper understanding of the essential aspects of brain activity. In Chapter 5, we examined the material structure of brain cells, specifically neurons and glia. Recall that a brain cell is made up of cytoplasm confined in a spatial region by the cell membrane consisting of a phospholipid bi-layer. There are various microorgans in the cytoplasm. Furthermore, specific protein molecules, called ion channels or membrane transport proteins, pierce the cell membrane and control the electric-charge-distribution difference between the interior and the exterior of the cell membrane by passing particular ions through them. Those membrane proteins provide each axonal neuron with a specific function as an electro-chemical circuit device of multi-input and single-output type with threshold-majority-decision logic. In other words, the axonal neurons as circuit devices of the neural network produce a single chemical output signal only when the sum of several inputs exceeds a certain threshold value, but there will be no reaction at all if this condition is not met. Neural networks in the brain are composed of numerous neurons with axons connected to each other by chemical synapses. Macroscopic changes of membrane-electric-potential difference, called neural impulses, realize the transmission of information by propagating in the neural network. Enfolding those neurons with axons, which form the neural network in the cerebral cortex, is a fine, extended structure called the dendritic network. The dendritic network, which is a dense network structure of dendrites of neurons with and without axons, might bear the highly advanced functions of memory and consciousness in the human brain. Here, we reach the front line of research in traditional brain science. Now we need the help of quantum theory to explore more deeply the sub-microscopic world of the brain. We have equipped ourselves with the fundamentals of **quantum mechanics** and quantum field theory to prepare for this expedition.

Let us revisit the cytoplasm of a brain cell such as a neuron or a glial cell. As shown before, cytoplasm is a complicated mixture of water (about seventy percent), protein (about twenty percent), and various kinds of **molecules** (about ten percent). A variety of microorgans manifest functional differentiation in the cytoplasm. This is a view of cytoplasm on the scale seen through an optical microscope. A completely different structure will appear if we look at cytoplasm on a much smaller scale, which we are incapable of examining with the optical microscope. At this deeper level, we find immediately beneath the cell membrane long, thin, filamentous strings made of helical chains of specific protein molecules of dumbbell-like form spanning several nanometers. The specific protein is called actin, and the filamentous string of actin is called an actin filament. Those actin filaments are connected to each other, and they form a dense network structure. They are, in part, also linked to specific membrane proteins embedded in the cell membrane. The spatial form of the cell membrane cannot be sustained without this dense network structure spanning the area immediately beneath the cell membrane. Furthermore, filamentous strings made of other proteins, called myosin, are linked to the actin filaments; they are called myosin filaments. Those actin and myosin filaments expand and contract as they withhold and release energy stored in the molecules of adenosine triphosphate (ATP). The movement of the cell membrane is realized by the spatial displacements of actin and myosin filaments, and, as a result, a biological cell, such as a neuron, becomes capable of performing amoeba-like movements. That is called an ATP cyclic process of the actin-myosin contraction and relaxation reaction, and it is known to be controlled by calcium ions.

The dense network structure of long, thin, filamentous strings of proteins is not restricted to that of actin and myosin filaments immediately beneath the cell membrane. There exists a vast, dense, three-dimensional network structure that extends throughout the cytoplasm. It is called a cytoskeletal structure or cytoskeleton, though it is not a static structure with a fixed geometrical configuration. It is dynamical. Within the microscopic world of the cell membrane, there appears to exist a dense, fine, dynamical network structure in which long, thin, filamentous strings of protein molecules attach to and detach from each other. The dynamical character of this network structure results from the deformation caused by attachable and detachable filamentous strings. The cytoplasm of a cell, in turn, manifests fluidity or deformability. There is an interesting parallel here between this microscopic structure and one described in cosmology: Long, thin, filamentous strings called cosmic strings spanned a dynamical, three-dimensional network structure throughout the whole universe until about ten thousand years after its creation. Those cosmic strings, assumed in quantum field theory to have been in a collective mode manifesting superconductivity, are thought to have transmitted energy as well as electric charge without thermal loss. Of course, that is the story in recent cosmology research.

Similar stories were proposed in 1968 and 1979 for the filamentous strings of proteins spanning the dynamical three-dimensional structure of the cytoskeleton in cytoplasm (Fröhlich 1968; 1970; Davydov 1979). Of course, even before ideas about the collective mode of filamentous strings of proteins were proposed by others, Umezawa had acquired a deep theoretical insight into the fundamental process of brain activity (Ricciardi and Umezawa 1967). Intrigued by the mysterious characteristics of memory, he had speculated that the essential aspect of the memory mechanism might be understood through a quantum field theoretical approach.

Fröhlich proposed a theory in which biomolecules with higher **electric dipole moment** line up along the actin filaments immediately beneath the cell membrane, and electric dipole oscillations propagate along each filament as coherent waves without thermal loss, just as in **superconducting media**. Such electric dipole oscillations are maintained by electrons trapped in and moving along the chain of protein molecules of each filament. This coherent wave is called a Fröhlich wave. According to Fröhlich's estimation, the **frequency** of a Fröhlich wave is in the region between 10^{11} and 10^{12} hertz, and it is called the Fröhlich frequency. Those Fröhlich waves, seen as energy quanta carrying energy given by the Fröhlich frequency times the **Planck constant**, can exchange energy with the electromagnetic field by creating and annihilating photons, that is, energy quanta of the electromagnetic field, corresponding to infrared light. Experimental evidence of Fröhlich waves has been obtained by means of microwave and infrared photon spectroscopy (Genberg et al. 1991; Genzel et al. 1976; Webb and Stoneham 1977; Webb 1980).

Davydov (1979) showed that superconducting electric current can be realized along the filamentous strings of protein molecules in the cytoskeletal structure by means of specific solitary waves carrying electric charge. This solitary wave is called a soliton. The soliton is a solitary wave that maintains its distinct form while in motion. Although it appears to exist independently, the soliton emerges from and sinks back into an underlying ground; the soliton is an expression of global properties.

Umezawa was the first to recognize the need for certain quantum theoretical degrees of freedom within the sub-microscopic world of the brain in order to explain its highly advanced functions, consciousness and memory specifically, and he looked for them inside the neuron (Ricciardi and Umezawa 1967). Later Fröhlich and Davydov presented plausible evidence of the existence of certain degrees of freedom, as described by quantum theory, along the filamentous strings of particular proteins. Then, after transferring to Edmonton, Canada,

Umezawa, in collaboration with Stuart and Takahashi, presented an epoch-making viewpoint, one radically different from that of the neuron doctrine originated by Waldeyer and Cajal at the end of the nineteenth century. From this new perspective, the conventional materialistic structure of the neuron, including the dense, microscopic network of the cytoskeletal structure enclosed by the membrane lipid bi-layer, was regarded as a superficial form of neuron. The essential structure of the neuron, considered to be the fundamental constituent of the brain realizing memory and consciousness, was thought to extend from the interior to the exterior of the neuron and to penetrate all the brain cells, including, of course, the glial cells. *Thus, the most fundamental unit of brain functioning—the quantum theoretical degrees of freedom—manifests spatial extension beyond a single cell.* Umezawa, Stuart, and Takahashi had predicted the existence of sub-microscopic degrees of freedom lying undiscovered behind the conventional materialistic structure of neurons and glia, degrees of freedom that extend to all the brain cells, which could be grasped only by quantum theoretical considerations. They speculated that such quantum theoretical degrees of freedom might be given by a field entity described by quantum field theory. They called that field entity a *cortical field.*

What is the cortical field in reality? In the quantum field theory of the brain developed by Umezawa, Stuart, and Takahashi, the dynamics of the cortical field at the sub-microscopic level is not isolated; it interacts with the macroscopic dynamics of the neural network. There the macroscopic dynamics of the neural network plays a merely pragmatic role in transmitting macroscopic signals to bodily tissues. How is it possible for the cortical field to interact with the conventional materialistic structure of the neurons? Those two questions might be strongly related. The latter provides a significant clue to surpass and replace the neuron doctrine with truly quantum theoretical conceptions.

We have just seen that the network of protein filaments forms a dense, three-dimensional, cytoskeletal structure extending throughout the cytoplasm of a brain cell. As a matter of fact, such filamentous strings of protein molecules are known to exist both inside and outside the cell membrane. There are many actin filaments inside and immediately beneath the cell membrane. There also exist many filamentous strings of particular proteins called collagens immediately adjacent to and outside the cell membrane. Those protein filaments existing outside the cell membrane form a fine network structure called an extracellular matrix, and a part of it is linked to specific membrane proteins embedded in the cell membrane. This network structure is similar to the dynamical network of protein filaments in the cytoskeletal structure. The most important point to note here is that *membrane proteins are linked simultaneously to the cytoskeleton inside and to the extracellular matrix outside the cell membrane.*

If, for the moment, we forget the existence of the conventional material structure of the cell membrane and the cytoplasm, we could discern a vast, microscopic, three-dimensional network structure spanning all the brain tissue by patching together the cytoskeletons and extracellular matrices. Furthermore, Fröhlich waves as well as solitons are propagating everywhere along the protein filaments in this vast microscopic network without loss of energy due to thermal disorder. These waves are elements of the ordered dynamics of the molecular vibrational fields of proteins and other biomolecules with high electric dipole moments along the filamentous strings described by collective modes in quantum field theory. The Fröhlich waves are produced by energy or charge stored in ATP molecules at specific membrane protein sites controlled mainly by calcium ions. Conversely, these waves of electric dipole oscillations can affect the electric-potential-sensitive, that is, charge-sensitive, ion channels embedded in cell membranes provided that the filamentous strings are linked with those ion channels. Those ion channels, affected by the Fröhlich waves, trigger microscopic physical processes of ionic diffusion and interface between the sub-microscopic dynamics of Fröhlich waves and the macroscopic dynamics of neural impulses. As a result, cooperative wave motions of the molecular vibrational field—for instance, the Fröhlich wave—may control the macroscopic dynamics of the conventional material structure of the neurons by means of membrane-electric-potential differences and ionic diffusions across the cell membrane.

Here we might be tempted to recall the wave cybernetics of the dendritic networks discussed in the preceding chapter. There, we used the theoretical framework of wave cybernetics subject to the optimal control principle. We developed a control theoretical formulation of the cybernetics of the dendritic network in terms of the collective physical phenomena taking place in the perimembranous regions immediately adjacent to the dendritic membrane. That control theoretical formulation is called wave cybernetics. We found that the cybernetics of the dendritic network could be described by a complex-number-valued order parameter subject to a Schrödinger-like wave equation. The wave cybernetics of the perimembranous regions immediately adjacent to the dendritic membrane control and affect the macroscopic activity of the neural networks of neurons with axons. The wave cybernetics of the dendritic network supposedly influence or modify the macroscopic cybernetics of the neural network.

Is the ordered and coherent propagation of Fröhlich waves and solitons nothing but the wave motion of the cortical field sought by Umezawa, Stuart, and Takahashi? That is, can the cortical field in the quantum field theory of the brain be identified with the molecular vibrational field of dipolar oscillations distributed along the filamentous strings of protein molecules in the vast, microscopic, three-dimensional network structure that spans all brain tissue by linking all the cytoskeleton and extracellular matrices? The answer is negative. Umezawa's cortical field lies concealed behind this global, microscopic, microphysical but still apparent material structure. Before proceeding to find that cortical field, let us consider the relationship between the wave cybernetics of the dendritic network and the quantum theory of the collective modes of molecular vibrational fields. This will help us to understand the way in which the sub-microscopic quantum dynamics in brain cells can be affected by the microscopic or macroscopic physical phenomena taking place in the dendritic network or the neural network.

Section 45
The Macroscopic Wavefunction as an Order Parameter

In the preceding chapter, we saw that the microscopic, cooperative physical phenomena taking place in the perimembranous regions immediately adjacent to the dendritic membrane may be described phenomenologically within the control theoretical framework of wave cybernetics. Within that framework, the cybernetics of the dendritic network is governed by the complex-number-valued order parameter given by the Schrödinger-like wave equation. There, we looked for a phenomenological description of the cybernetics of the dendritic network based on the optimization principle of cybernetics.

In the preceding section, we showed that there are densely packed network structures composed of filamentous strings of particular protein molecules immediately adjacent to the cell membrane of brain cells such as neurons and glia. Furthermore, there exists ordered dynamics of molecular vibrational fields of proteins and other biomolecules with high electric dipole moment along those filamentous strings. Ordered dynamics is characteristic of collective modes in quantum field theory. There, the creation and annihilation dynamics of energy quanta realizes a long-range coherence or harmony so that the dynamics manifests a macroscopic order: The electric dipole moments of molecules of filamentous strings manifest harmonic oscillations, and their energy quanta are in the same energy eigenstate described by the wavefunction within a macroscopic spatial region comparable in size to the coherence length. Those energy quanta in a macroscopic region in the same energy eigenstate can transfer energy collectively to another energy eigenstate described again by another wavefunction; therefore, they are said to be in the collective mode. The collective mode can be illustrated by a wavefunction taking a uniform functional form in the macroscopic region, that is, a macroscopic wavefunction. Recall that the time development of the macroscopic wavefunction is described by the macroscopic Schrödinger equation, as shown in Chapter 3.

We obtain an interesting physical picture of the cooperative, microscopic phenomena taking place inside and outside the cell in the regions immediately adjacent to the cell membrane, that is, in the perimembranous regions. Those phenomena are described as collective modes in quantum field theory and are called macroscopic quantum phenomena. They are sub-microscopic, ordered dynamical phenomena that appear even on a macroscopic scale, though they are still subject to the principles of quantum theory. Emergence of sub-microscopic, ordered dynamics on a macroscopic scale is ensured by the universal effect known as *spontaneous symmetry-breaking* in quantum field theory (Umezawa 1993). It will be explained in Chapter 8. Similar to other typical physical phenomena of collective modes, such as superconductivity and lasers, those macroscopic quantum phenomena can be described by macroscopic wavefunctions. Indeed, energy as well as electric charge can be transferred in the macroscopic region of the collective mode without loss due to thermal dissipation and fluctuation, as evidenced in a **superconducting medium**.

In this sense, the perimembranous region realizing the collective mode can be regarded as a macroscopic superconducting region in which energy as well as electric charge transfer without loss due to thermal dissipation and fluctuation. The important physical feature of the cell membrane might be described as a sandwich structure with an insulating lipid bi-layer between the macroscopic superconducting regions of the collective modes of molecular vibrational fields. Such a sandwich structure of superconducting and insulating layers in non-living matter is known as a Josephson junction. The Josephson junction manifests a macroscopic quantum effect in which superconducting electric current across the sandwich structure can be rigorously controlled by the electric potential difference imposed between the upper and lower superconducting layers. Contrary to ordinary matter, the electric current vanishes if a constant electric potential difference is imposed, and it appears if an alternating electric potential difference is imposed. The strange effect of the Josephson junction was first predicted theoretically by **Brian Josephson**. In 1973 he received the Nobel Prize for Physics for that prediction.

We have revealed a new microscopic structure of the cell membrane: Some specific regions of the cell membrane form a sandwich structure with an insulating lipid bi-layer between the macroscopic superconducting regions of the collective modes of molecular vibrational fields in the perimembranous regions. They are nothing else but Josephson junctions, which may be called Josephson biojunctions. The electric current flowing across the Josephson junction, realized by sub-microscopic quantum dynamics, is very sensitive to the macroscopic electric potential differences associated with the **Josephson effect**. It is this sensitivity of the sub-microscopic quantum dynamics that enables the microscopic or macroscopic physical phenomena taking place in the dendritic network or the neural network to affect the sub-microscopic quantum dynamics in the brain cells.

The microscopic material structure of the cell membrane reveals that the dendritic network of membranes may be regarded as a fine and sophisticated electric circuit of extremely many Josephson junctions, which could be likened to a superconducting computer or a Josephson computer. The cybernetics of that huge electric circuit can be described well by the macroscopic wavefunction subject to the macroscopic Schrödinger equation. There is now justification for identifying the order parameter and the Schrödinger-like wave equation governing the wave cybernetics of the dendritic network with the macroscopic wavefunction and the macroscopic Schrödinger equation, respectively. The phenomenological description within the framework of optimal control theory, developed in Chapter 6, is now supported by the collective mode in quantum field theory.

Now we can interpret the wave cybernetics of the dendritic network in terms of the macroscopic wavefunction governed by the macroscopic Schrödinger equation characteristic of the ordered dynamics of the collective mode in quantum field theory. Hence, the collective modes of molecular vibrational fields along the filamentous strings of protein molecules in the perimembranous regions immediately adjacent to the cell membrane might facilitate the highly advanced functions of the brain, such as memory and consciousness. However, since the microscopic material structure is not much different from a superconducting computer device equipped with a huge electric circuit of Josephson junctions, the cybernetics is essentially the same as that of the neural network, though the signal speed increases and the energy consumption decreases. Even if we introduce the molecular vibrational fields along the protein filaments in the dense network structures of the perimembranous regions immediately adjacent to the cell membrane, the highly advanced activity of the brain might not be derived from them. They might simply provide a dense superconducting network of Josephson junctions that directly affect the macroscopic cybernetics of the neural network. Therefore, they cannot be identified with Umezawa's cortical field. Instead those molecular vibrational fields along the protein filaments in the dense network structures of the perimembranous regions immediately adjacent to the cell membrane seem to provide a physical interface between the cortical field and the conventional neural network.

Section 46
The Corticon and Water

The wave motion of any field entity in the sub-microscopic world can be described within the realm of quantum field theory. The wave motion of the electromagnetic field with a definite **wave number** can be regarded as energy quanta, called photons, which move as particles do. The same might also apply to any wave motion of the cortical field extending to all the brain cells of the cerebral cortex, though the reality of this has not yet been confirmed. According to the basics of quantum field theory, the wave motion of these undiscovered, sub-microscopic degrees of freedom of the cortical field might be thought of as energy quanta behaving as particles. Stuart, Takahashi, and Umezawa (1978; 1979) called this energy quantum a *corticon*. Corticons exist both inside and outside neurons and glia everywhere in the cerebral cortex because the cortical field extends through the whole cortex. Of course, corticons might also exist in other parts of the brain tissue, but we are restricting our discussion to the cerebral cortex since we are mainly concerned with the mechanism of the highly advanced functioning of the human brain evident in memory and consciousness. The cerebral cortex covering the outer layer of brain tissue can be imagined as a cryptic universe in which energy quanta called corticons undergo the creation and annihilation dynamics described by quantum field theory.

It was Stuart, Takahashi, and Umezawa's belief that nothing but the creation and annihilation dynamics of corticons could give rise to the fundamental physical process that serves as the mechanism of memory and consciousness in the cerebral cortex. Once a vacuum state of the cortical field in the sub-microscopic world of the brain is regarded as memory, then both the infinite variety and the imperfect stability of memory might be explained within the realm of quantum field theory. Furthermore, it can be concluded that memory manifests nonlocality because the vacuum state of a certain field entity is regarded as a stable physical object with macroscopic spatial extension. *Consciousness is then nothing but the creation and annihilation dynamics of corticons, that is, energy quanta of the cortical field, realized in the vacuum state representing memory.*

However, neither the corticon nor the cortical field could be found as realistic entities when Stuart, Takahashi, and Umezawa developed their theoretical framework of the quantum field theory of the brain. In any case, the interface between the fundamental process of brain functioning in terms of the creation and annihilation dynamics of corticons and the conventional cybernetics of neural networks occurs in the collective modes of molecular vibrational fields along filamentous strings of protein molecules spanning an extensive, dense, three-dimensional network structure that connects all the cytoskeletons and the extracellular matrices of the brain cells in the cerebral cortex. Those collective modes are realized in the coherent and cooperative dynamics of many electric dipoles, such as Fröhlich waves and Davydov solitons. Therefore, the cortical field as well as the corticon might be capable of interacting with those collective modes of the molecular vibrational fields of electric dipoles. This fact provides a clue to discovering the yet unknown physical degrees of freedom of the cortical field and the corticon.

What other candidate is there for the cortical field, besides the microscopic, material structure of the cytoskeletons and the extracellular network of filamentous strings of protein molecules? As far as typical material constituents are concerned, there are no other known microscopic structures either inside or outside the brain cells. The brain cells are all made of numerous species of biomolecules—such as proteins, lipids, and so on—that exchange atoms, ions, and electrons among themselves. There is nothing more. Is there room for the cortical field in the fundamental, material structure of the brain cells on the submicroscopic scale? Yes, there is certainly plenty of room for the cortical field.

It seems rather interesting that in the conventional picture of a brain cell, such as a neuron or a glial cell, obtained in molecular biology, one is inclined to liken the material structure of the cell to the architecture of man-made buildings. Filamentous strings of proteins that form the cytoskeleton are regarded as beams and uprights, and lipid bi-layers are seen as walls that divide the whole building into separate compartments. Although it is well known that more than seventy percent of the material composition of a brain cell is water, its essential importance is usually dismissed. That is, water is viewed merely as a structureless, space-filling, background medium in which molecular biological and biochemical events occur. Of course, hydrophilic and hydrophobic effects as well as hydrolysis reactions are obviously passive roles for water in biological cells in general. Until recently no active roles for water have been acknowledged.

It seems worthwhile to note here that the essential importance of water in the living cell has not been completely dismissed in biology. For more than twenty years, John Watterson in Australia has been tackling the difficult issue of the role of water (Watterson 1987). In a radically new model of water, developed by Watterson through biological considerations, water is an integral component in the microscopic, material structure of all cellular assemblies, and it actively participates in their functions. Despite Watterson's long-standing endeavor to promote the importance of water in the fundamental processes of biological cells, water has remained outside the consideration of modern molecular biology simply because water seems to be a trivial material component of living matter. This is certainly not the case, however, in Stuart, Takahashi, and Umezawa's quantum field theory of the brain. Though almost everyone else has dismissed the active role of water, its essential importance to higher brain functions, such as memory and consciousness, was recognized in the original idea of the cortical field and corticons proposed by Stuart, Takahashi, and Umezawa. *Indeed, water is not a trivial material component of living matter, but the most fundamental one from which all the basic aspects of life emerge.*

Water is a typical molecule consisting of one oxygen atom and two hydrogen atoms, simple in its form but rich in its physical characteristics. Its richness is derived from the simplicity of its form. That is, due to the spatial, geometric configuration of the two hydrogen atoms relative to the single oxygen atom, the water molecule has a constant electric dipole moment. The water molecule has a symmetrical form invariant under reflection with respect to a fictitious mirror situated just between the two hydrogen atoms. To get an intuitive image of the form of a water molecule, imagine a figure of Mickey Mouse. The face of Mickey Mouse stands for the oxygen atom of the water molecule, and his two ears stand for the two hydrogen atoms of the water molecule. In this whimsical sketch, the constant electric dipole of a water molecule can be shown as an arrow of constant length pointing from the top of Mickey Mouse's head to the center of his neck. This arrow is a vector that represents the constant electric dipole moment characteristic of the water molecule. The axis of the arrow is the symmetry axis of the water molecule.

The water molecule has several degrees of freedom: it can move freely in space, rotate freely in space, and manifest molecular vibrations. (Mickey Mouse's face can move freely on the screen, rotate freely on the screen, and his ears can vibrate!) Among these degrees of freedom, the most essential and important one for water is the rotation around the symmetry axis of the water molecule, that is, the constant electric dipole moment of the water molecule. Thus, each water molecule can be regarded as a quantum mechanical spinning top with a constant electric dipole moment. (Imagine a figure of Mickey Mouse revolving rapidly around the arrow of constant length.) A quantum mechanical spinning top with a constant electric dipole moment interacts with the collective mode of molecular vibrational fields of biomolecules with higher electric dipole moments. That is so simply because both the former and the latter can be regarded as electric dipoles engaged in electromagnetic interaction with each other. Therefore, the totality of the quantum theoretical, rotational degrees of freedom of all available water molecules, seen as a field entity that extends through the whole assembly of brain cells, might be the best candidate for the cortical field. In quantum field theory, the totality of spinning tops distributed in a spatial region can be well described by the notion of a **spinor field**. We will call it a water rotational field or a water dipole field.

Physical inspection leads us to believe that the cortical field hypothesized in Stuart, Takahashi, and Umezawa's quantum field theory of the brain could be realized as a water rotational field extending over the whole assembly of brain cells (Jibu and Yasue 1992; 1993A; 1993B; 1993C). *It is astonishing that water, which has been tacitly assumed to play only a trivial role, is now highlighted as the central material constituent participating in the fundamental physical phenomena that evoke highly advanced brain activities, such as memory and consciousness.* Contrary to its apparent triviality, as suggested by its familiar, universal presence in the life of our planet, water is not a simple entity. It might be of interest, therefore, to investigate the basic physical properties of water before we present a quantum field theory of the brain that cites the water rotational field as the cortical field.

Section 47
Water and Life

To the best of our knowledge, no living matter with biological cells, from prehistoric times to the present, has maintained life without water. Water must be the most essential and important material constituent of living matter, though its role has been underestimated to the point that it is regarded merely as a space-filling, background medium for molecular biological processes. Water is so common to living matter that its active physical properties have been tacitly assumed to be inconsequential. However, the true nature of water far exceeds the naive and superficial understanding of it arising from our familiar daily experience. It is no exaggeration to say that nobody fully understands what water is. Generally, people know that water is made of numerous water molecules, H_2Os. However, this seemingly elementary fact cannot be justified from the point of view of quantum theory. The truth is that water is an extremely large, single molecule composed of many fundamental units, each of which consists of an oxygen atom and two hydrogen atoms. That is, each unit of H_2O is bonded to other units depending on the dynamics of the electrons in the energy eigenstates of the atoms (Franks 1972).

Two atoms or two molecules are thought to be bonded to each other if at least one electron occupies an energy eigenstate of one atom or molecule and simultaneously occupies an energy eigenstate of another atom or molecule. (Recall that the energy eigenstate of an atom or molecule is the dynamical state of an electron moving around the **nucleus** with a particular value of energy known as the energy eigenvalue. Both the energy eigenstate and the energy eigenvalue are obtained by solving the stationary Schrödinger equation.) It might be helpful to imagine a couple (a man and a woman) having at least one child. The man and woman are bonded by the existence of the child, who occupies the position of a son or a daughter of the man and simultaneously occupies the position of a son or daughter of the woman. Similarly, two hydrogen atoms are thought to be bonded to each other if at least one electron is simultaneously occupying the energy eigenstates of two hydrogen atoms. The two hydrogen atoms thus bonded to each other come to be a hydrogen molecule. This bonding mechanism of atoms and molecules, peculiar to quantum theory, is called covalent bonding, and it applies not only to hydrogen atoms but to other atoms as well. This covalent bonding enriches the physical structure of molecules by bonding several specific atoms together.

The mechanism of covalent bonding can be understood in terms of the total energy of the electrons. The total energy of all the electrons in two atoms becomes lower if the two atoms are bonded to each other by covalent bonding to form a molecule, that is, if some of those electrons happen to occupy certain energy eigenstates of one atom and simultaneously occupy certain energy eigenstates of another atom. This means that a certain amount of energy is needed to break a molecule into two separate atoms. The lower the total energy of a physical system, the greater the likelihood that this physical system will be realized in the natural world. Therefore, there are more molecules than atoms in the natural world.

A hydrogen atom, once bonded to a certain atom by covalent bonding and formed a molecule, can be still bonded to another atom or molecule by the electric attraction of Coulomb force between positive and negative charges: As an electron of the hydrogen atom is half belonging to the other atom, the hydrogen atom loses negative charge and so becomes positively charged. This bonding mechanism of molecules, realized by covalent bonding of an hydrogen atom and the electric attraction of Coulomb force, is called hydrogen bonding or simply H-bonding.

Water might be the biggest molecule in the world. All the oxygen and hydrogen atoms in water are thought to be bonded to each other by hydrogen bonding. The totality of water on the planet Earth could be considered one huge water molecule. (Of course, this might be an overstatement, but it helps us to understand the most important aspect of water.) Each unit configuration H_2O, which used to be regarded as a water molecule composed of an oxygen atom and two hydrogen atoms, can no longer be identified with a water molecule but simply a fundamental unit of atoms strongly bonded to each other. However, at the bottom of the oxygen atom, that is, at the neck of Mickey Mouse, each unit configuration H_2O can still be bonded to a hydrogen atom of another unit by hydrogen bonding. (The neck of Mickey Mouse can be bonded to an ear of another Mickey Mouse whose neck can also be bonded to the ear of a third Mickey Mouse, and so on.) As a result, all the H_2O units that are hydrogen bonded create a huge water molecule. Tremendous energy is required to break this single, large water molecule into separate H_2O units. That might explain the high specific heat of water in contrast to other kinds of matter. The bonding energy between H_2O units is actually small, that is, of a magnitude routinely available in thermal energy, so the superstructure of this huge molecule of water is in constant flux, with the intermolecular bonds constantly being broken and reformed. Thus, the huge water molecule may not be understood literally as a steady, solid molecular structure but as a dynamic one subject to constant breakdown and restoration of the intermolecular bonds.

Each H_2O unit has a proper electric dipole moment due to the fact that the two hydrogen atoms are both bonded to the upper region of an oxygen atom, just as Mickey's two ears are both bonded to the upper region of his face. Covalent bonding between an oxygen atom and a hydrogen atom is established by the sharing of common electrons, so the **probability** distribution of electrons around the oxygen atom is shifted to the hydrogen side. Then, the hydrogen side of the oxygen atom, that is, the crown of Mickey's head, becomes negatively charged due to the presence of electrons with negative electric charge. The opposite side of the oxygen atom, that is, Mickey's neck, becomes positively charged because of the relative absence of electrons with negative electric charge. That is an effect of covalent bonding. The electric dipole of each H_2O unit manifests a strong sensitivity to the electric dipoles of other units due to the electromagnetic interaction between two electric dipoles, thus realizing another binding mechanism different from covalent bonding, that is, hydrogen bonding. This binding mechanism is of a purely electromagnetic nature; it emerges not only at the quantum theoretical level but also at the classical level as a dipole-dipole interaction. It enables the units of H_2O to join and to maintain a certain structural configuration among themselves. Furthermore, any other kinds of molecules with constant electric dipole moment can be joined in this structural configuration. Biomolecules, such as proteins and nucleic acids, with constant electric dipole moments gather and connect to the water units.

A more radical idea is that the structural configuration of water gathers specific biomolecules with electric dipole moments so that, in a sense, they fit the structural configuration of water itself. That is, a single, large water molecule, which is composed of numerous units of H_2O bonded together, includes specific biomolecules within its combined structure. The arrangement of the bonding structures of those biomolecules adjusts to the overall structural configuration of water. In this case, the inconspicuous water, not the apparent biomolecules, initiates the structural configuration essential to living matter.

We should always keep in mind that more than seventy percent of the human body is water—a single, huge, cryptic water molecule rich in sub-microscopic physical structure. This fact might force us to change our view of life in a radical way. *Perhaps life is nothing but the unity of water as a single molecule in living matter.* Matter previously regarded as "living" is said to become dead or non-living if the unity of water in that matter breaks down. That is, a single, huge water molecule rich in sub-microscopic physical structure in living matter breaks into a less structured assembly of H_2O molecules when that living matter dies. In the next chapter, we will develop a purely scientific approach to the two remaining questions in basic research: "What is **mind**?" and "What is life?" We will address those two issues from the new point of view that *water is the fundamental constituent of both life and consciousness.*

Some of you might have read the novel *SOLARIS* written by Stanislaw Lem (Lem 1961). It was also made into a movie. Here is the story line:

> There once was a planet called Solaris. It was a peculiar planet made of water. There was no land, only sea. Scientists who were in a space ship orbiting this sea planet began to suffer from peculiar personal experiences. All of the scientists met and talked with de-

> ceased friends or relatives who existed only in their memories. They were all thrown into confusion, and earnest inspection was needed. They finally found that the planet Solaris was indeed living. The "sea" manifested a systematic, orderly flow, and it thought by its flow, the pattern of which was directed towards a precise objective. The sea knew the minds and memories of the scientists. It tried to create communication between them and the deceased by reproducing images of their friends and relatives in the scientists' minds through a holographic process.

The story of *SOLARIS* seems highly suggestive. Perhaps it foreshadows a drastic and fundamental change in our point of view regarding the basis of life. (Much of science fiction does seem prophetic.) The planet Solaris might represent a single, gigantic water molecule that manifests an overall unity. By our new definition of life, Solaris would indeed be living.

Section 48
The Exchange Boson and the Photon

In Stuart, Takahashi, and Umezawa's quantum field theory of the brain, quanta of the cortical field, called corticons, undergo creation and annihilation dynamics in the cryptic universe of the brain, just as quanta of the **matter field** do in our universe. *The fundamental processes of the brain that give rise to the essential mechanism of memory and consciousness might be nothing but quantum theoretical phenomena caused by the creation and annihilation dynamics of corticons.* Then, what kind of phenomena are they? In any physical phenomenon in our universe caused by the dynamics of electrons, photons play an important role. One example is the aurora, which is nothing but the creation of photons by electrons moving in the ionosphere. Another example is the fire ball created by a high energy, electron-**positron** collision. *As a matter of fact, photons and electrons are closely related to each other. Any two electrons can affect each other in such a way that one electron emits (i.e., creates) a photon and another electron absorbs (i.e., annihilates) this photon.* The close relationship between electrons and photons is revealed in the interaction between the wave motion of the matter field of the electron, that is, its probability amplitude, and the wave motion of the electromagnetic field within the framework of quantum field theory.

Similarly, nucleons (i.e., protons and neutrons) and mesons are closely related to each other. Any two nucleons can affect each other in such a way that one nucleon emits (creates) a meson and another one absorbs (annihilates) that meson. This is what Yukawa discovered in 1935. *Electrons can manifest various physical phenomena in our universe only when photons are present. According to quantum field theory, nothing could happen in the universe if there existed only electrons.*

The same argument certainly applies in the case of corticons. There would be no meaningful physical phenomena in the sub-microscopic world of the cerebral cortex if only corticons existed there. Thus, Stuart, Takahashi, and Umezawa looked for another field entity in the cerebral cortex, one corresponding to the electromagnetic field of photons evident in physical phenomena realized by the dynamics of electrons. They assumed this field to be a Bose field and called it simply a Bose field without giving it a more specific name. At the time that Umezawa, in collaboration with Stuart and Takahashi, proposed the idea of a Bose field interacting with the cortical field, the physical reality of the Bose field was not known (Stuart et al. 1978; 1979). It was farsighted that they did not give it a specific name. As will soon be shown, this Bose field turned out to be an already well-known Bose field!

If we apply quantum field theory to the Bose field, its wave motion can be regarded as a particle motion of an energy quantum, just as in the case of the electromagnetic field. Quanta of the electromagnetic field are called photons, and those of the Bose field in question are called exchange bosons. Any two corticons can affect each other in such a way that one corticon creates an exchange boson and the other one annihilates the exchange boson. Thus, the fundamental processes of memory and consciousness might arise as quantum theoretical phenomena in which corticons and exchange bosons are subject to sophisticated creation and annihilation dynamics. In other words, *the fundamental processes of the brain could be nothing other than the interaction between corticons and exchange bosons.*

We have finally arrived at a place from which the entire picture of the submicroscopic physical world of the brain can be seen. It is the front line of brain science first attained by the intuitive power of Umezawa. From here we can clearly envision a cryptic universe sparkling with the creation and annihilation dynamics of corticons and exchange bosons. You might be eager to see with your own eyes the way in which memory is realized. You might want to know how memory emerges in consciousness, why some memories are associated with each other, and why certain memories are lost. However, it seems worthwhile to investigate here the Bose field and the exchange bosons. The reason is that we need to acquire a realistic image of them as concrete physical entities. The cortical field has been identified with the water rotational field, which is nothing but the totality of the many quantum theoretical rotational degrees of freedom of H_2O molecules extended over the whole assembly of brain cells. More concretely, corticons can be grasped as energy quanta of the water rotational field in the brain. Is it possible to devise a realistic picture of the Bose field, that is, a concrete image of the exchange boson interacting with the water rotational field and its energy quanta?

Corticons are supposed to interact with each other by emitting and absorbing exchange bosons. Corticons are nothing but energy quanta of the water rotational field whose degrees of freedom are rotations of water molecule units (H_2Os) around their symmetry axes aligned with their proper electric dipoles. In other words, the cortical field can also be regarded as an electric dipole field extending to the whole assembly of brain cells in the cerebral cortex. This is so simply because each H_2O unit can be considered a quantum mechanical spinning top with constant electric dipole moment. The electric dipole, composed of a positive electric charge and a negative electric charge slightly separated in space, interacts as a paired unit with the electromagnetic field. In that sense, it may be natural to call the water rotational field "the water dipole field." Thus, we see immediately that the cortical field must manifest an interaction with the electromagnetic field. *It is likely, therefore, that the Bose field interacting with the cortical field would be the electromagnetic field. Exchange bosons created and annihilated by corticons, that is, energy quanta of the water rotational field, are now found to be photons.*

Just as electrons interact with each other by creating and annihilating photons, the creation and annihilation dynamics of photons and energy quanta of the water rotational field may manifest an order characteristic of quantum field theory. What kind of order is it? And what sort of mechanism might be found in the brain by investigating the order in the creation and annihilation dynamics of photons and energy quanta of the water rotational field within the realm of quantum field theory?

Chapter 8
Quantum Brain Dynamics

Section 49
QBD

The **quantum field theory** of the **matter field** of the **electron** and the **electromagnetic field**, which describes all physical phenomena involving electrons and **photons**, is called *quantum electrodynamics* and abbreviated to QED. Analogously, the quantum field theory describing the fundamental processes of the strong interaction between nucleons and mesons in terms of quarks and gluons is called *quantum chromodynamics* and abbreviated to QCD. In this theory, quarks interact with gluons in a manner analogous to electrons interacting with photons, and sophisticated, high **energy**, physical phenomena are thought to be realized in the creation and annihilation dynamics of quarks and gluons. Delving more deeply into the sub-microscopic and sub-nuclear realm, even the geometry of space and time falls into the scope of quantum field theory. This speculative framework has been called *quantum geometrodynamics* and abbreviated to QGD. Besides these, we have a quantum field theoretical framework to describe the creation process of our universe, called *quantum cosmology*, and another to investigate macroscopic thermodynamical phenomena, called *thermo field dynamics*. The last one was also originally developed by Umezawa and Takahashi (Umezawa 1993).

Thus, quantum field theory is really a very general and universal theoretical framework, which provides us with fundamental physical laws capable of expressing various physical phenomena in a unified way on the sub-microscopic, the microscopic, and the macroscopic scales. The fundamental processes in the sub-microscopic world of the brain that give rise to the mechanism of human **memory** and **consciousness** might be no exception. Following Stuart, Takahashi, and Umezawa (1978; 1979) along the path of quantum field theory, we are going to get a clear view of the sub-microscopic world of the brain. We should not be scolded if we call this completely new branch of quantum field theory *quantum brain dynamics* and abbreviate it to QBD.

We shall investigate the typical physical phenomena arising from the creation and annihilation dynamics of corticons, that is, energy **quanta** of the water rotational **field** extending to the whole assembly of brain cells, and photons, that is, energy quanta of the electromagnetic field. We hope to arrive at a better and deeper understanding of not only memory and consciousness but also of life. We will call the quantum system of the water rotational field and the electromagnetic field extending to the whole assembly of brain cells in the cerebral cortex the QBD system. It is not a closed system isolated from other physical systems, but an open system interacting with the surrounding systems. Since the QBD system is a quantum system described by **quantum mechanics** and quantum field theory, it can be understood to interact directly and mainly with other quantum systems surrounding the QBD system. They are called the external systems of the QBD system and include the following: the quantum system of **electric dipole** oscillations along the filamentous strings of proteins in the cytoskeletal structures and in the extracellular matrices of the brain cells and the macroscopic quantum system of charged elements in the perimembranous regions immediately adjacent to the cell membranes of the brain cells. Those external systems play important roles in the transmission of information between the QBD system and the macroscopic cybernetic system of the dendritic and neural networks.

There are two basic fields in QBD: the water rotational field and the electromagnetic field. The two must be described simultaneously by quantum field theory because they interact with each other. This is essentially the same as describing **wave** motions of the electromagnetic field in quantum field theory. However, since the two fields interact with each other and the water rotational field manifests a symmetry property (as will be seen later on), the minimum energy eigenstates of the wave motions of those two fields, that is, the vacuum states, become approximate. Although the electromagnetic field can be described exactly as it was in Chapter 3, the water rotational field needs further consideration.

In the quantum field theory of the electromagnetic field, the vacuum state is the dynamical state in which the length of the vector potential at each point in space fluctuates around zero, subject to a **probability** distribution representing quantum fluctuations. In QBD, similar dynamical states are thought to exist in the quantum field theory of the interaction between the water rotational field and the electromagnetic field. One of them, for example, might be the dynamical state in which the electric-dipole-moment **vector** (represented by the arrow through the figure of Mickey Mouse) of each H_2O **molecule** at each point in space in a macroscopic region of the brain fluctuates around a common, constant vector, subject to a probability distribution representing the quantum fluctuation. In other words, all the electric-dipole-moment vectors of H_2O molecules in this macroscopic region are aligned in one and the same direction and manifest random deviation from the aligned direction corresponding to quantum fluctuation.

Such a dynamical state is called a coherent state or classical vacuum in quantum field theory. It is considered to be an approximate vacuum state in the case of an interacting field, such as the water rotational field. A vacuum state of the interacting water rotational field and the electromagnetic field is given by the product of the vacuum state of the electromagnetic field and one of the approximate vacuum states of the water rotational field. We call this a QBD vacuum.

In a sense, a QBD vacuum is nothing but a collective mode of the interacting water rotational field and electromagnetic field. That is, the rotational **degrees of freedom** of the H_2O molecules all fall into the same energy eigenstate of minimum energy within a spatial region whose macroscopic size is given by the coherence length of the collective mode. There are infinitely many different QBD vacua because there are infinitely many energy eigenstates of minimum energy that correspond to the infinitely many possible directions of the H_2O molecule's electric-dipole-moment vector (i.e., the arrow through the figure of Mickey Mouse). Recall that the direction of the electric dipole can be chosen freely, while the length or strength of it is kept constant due to the specific asymmetrical configuration of the H_2O molecule. The coherence length of a QBD vacuum is estimated to be inversely proportional to the energy difference between the first excited state (i.e., the second lowest energy eigenstate) and the minimum energy eigenstate (i.e., the lowest energy eigenstate) of the rotational degrees of freedom of an H_2O molecule. Regarding the H_2O molecule as a quantum mechanical rotator or a top spinning around its electric dipole, this energy difference can be estimated immediately, obtaining about 200 cm^{-1}. The coherence length of a QBD vacuum, hence, becomes about 50 microns.

We obtain an interesting physical picture of the vacuum state of the interacting water rotational field and electromagnetic field in QBD. The water rotational field extending over the whole assembly of brain cells manifests an overall domain structure in which the whole region occupied by water is divided into many domains, each less than 50 microns in size. In each domain, all the H_2O molecules manifest a quantum mechanical coherence in the sense that they are in a specific collective mode in which their proper electric-dipole-moment vectors are all aligned in a uniform, constant direction. This uniform direction can be altered from domain to domain and can realize a different QBD vacuum in each domain.

Since all the H_2O molecules in a single QBD vacuum domain are in a collective mode, they behave uniformly, just as the members of a synchronized swimming team do. In other words, they are bound to each other quite strongly by a quantum theoretical nonlocal coherence. It is important to understand that in a collective mode, each H_2O molecule is subject to purely quantum theoretical **nonlocality**, which was energetically investigated by **David Bohm**. No longer applicable is the conventional causal point of view, in which each element interacts only with its nearest neighbors and a chain of local and causal interactions among all the elements governs the totality of elements. Every dynamical system in **classical physics** is subject to this causality, but this is not so in **quantum theory**. Imagine a team of swimmers, each of which observes and imitates the movement of the nearest teammate. The team might manifest all the movements, but its performance would be far from the best in *synchronized* swimming. Nonlocal coherence must be found in the movement of every swimmer if the team is to be a truly competent *synchronized* swimming team.

It is also interesting that all the H_2O molecules in a QBD vacuum domain are perceived as bonded to each other by quantum nonlocality or long-range coherence; this maintains order in their configuration. This nonlocal bonding mechanism, characteristic of the collective mode of quantum field theory, emerges even at the macroscopic scale and results in the quantum theoretical macroscopic ordered state familiar in **superconductivity** and lasers. Hence, all the H_2O molecules bound together in a QBD vacuum domain form a single, extensive molecule of water in a macroscopic domain. That is, water throughout the whole region of the cerebral cortex is thought to be composed of many macroscopic water molecules whose sizes are all comparable to the long-range coherence length of the QBD vacuum, that is, about 50 microns. This remarkable feature of water in living matter—for example, in brain tissue—might provide us with quantum field theoretical support for the radically new idea developed in Chapter 7; namely, that *life is nothing but the unity of water as a single molecule in living matter.* Water is made of many macroscopic water molecules that border each other in living matter, whereas it is composed of numerous separate H_2O molecules in non-living matter. The fictional planet Solaris would be the ideal case in which one of the macroscopic water molecules absorbs the other ones by enlarging its coherence length throughout the entire sea of Solaris. Indeed, the fact is much more unconventional than the fiction.

Section 50
Spontaneous Symmetry-Breaking and the Goldstone Boson

The vacuum state, that is, the lowest energy eigenstate of the water rotational field in QBD, manifests a rich dynamical structure composed of domains that border each other. In each domain, the H_2O molecules demonstrate quantum nonlocality. They are bound in such a way that their electric-dipole-moment vectors are aligned in one and the same direction up to quantum fluctuations. That is, the vectors as a whole generally point the same direction, but there is slight fluctuation in that the direction occasionally changes to a small degree.

Each domain has a macroscopic spatial extension on the scale of the coherence length, that is, 50 microns, and carries ordered dynamics in the collective mode of the H_2O molecules. The domain represents not only an approximate vacuum state of the water rotational field but also a macroscopic water molecule extended to the scale of the coherence length (50 microns). The vacuum state in QBD is given by the product of the lowest energy eigenstate of the water rotational field and that of the electromagnetic field. The vacuum state of the electromagnetic field manifests a single, simple structure and is uniquely determined as the lowest energy eigenstate. The vacuum state of the water rotational field, on the other hand, manifests infinite multiplicity or degeneracy because the aligned electric dipole moments of all the H_2O molecules in a domain can take any one of an infinite number of spatial directions. This means that in a single domain an infinite number of vacuum states are possibly realized.

Recalling our favorite image of the synchronized swimming team, we can illustrate the QBD vacuum state as follows: What is the lowest energy state, that is, the vacuum state, of the movement of a team performing synchronized swimming? Of course, any state in which all the swimmers stand still can be the lowest energy state. There are many possibilities: each swimmer can stand still and turn her face any direction she likes. In that case, the audience sees all the swimmers on the team facing different directions at random and cannot expect a good performance of synchronized swimming. It may be quite hard for the swimmers even to start their exhibition of synchronized swimming in such a disordered state. In that sense, the disordered state in which each swimmer stands still and turns her face any direction she chooses cannot be the lowest energy state of the movement of the synchronized swimming team.

Any state, including the lowest energy state, of a synchronized swimming team must manifest order or harmony so that the audience can see order or harmony in the performance. Therefore, the lowest energy state, that is, the vacuum state, of the motion of a synchronized swimming team is a state in which all the swimmers stand still and turn their faces one and the same direction, say, toward the front entrance of the hall. Then, the audience can see harmony and order even in the still state of the team. Each teammate can easily start a good performance of synchronized swimming from that still, ordered state.

Note that there is not only one vacuum state of the motion of a synchronized swimming team. There are actually an infinite number of vacuum states. Any state in which all the swimmers stand still with their faces turned the same direction can be a vacuum state, and there is no preferred choice of direction. A state in which all the teammates are facing the emergency exit is one vacuum state. Facing the press box, the swimmers are in another vacuum state. Similarly, there is no single vacuum state for the water rotational field. There are actually an infinite number of vacuum states.

In addition to the infinite number of vacuum states realized in QBD, there are also extremely many domains in the cerebral cortex, so the number of different combinations of aligned directions among those domains becomes infinitely large. Therefore, there exists an infinite number of vacuum states of the water rotational field. The vacuum state in QBD can be obtained simply by combining the vacuum state of the water rotational field with the vacuum state of the electromagnetic field. However, the vacuum state of the electromagnetic field is unique—there is only one—so we can disregard it hereafter when we specify the vacuum state in QBD. The rich domain structure of the vacuum state of the water rotational field is assigned to the QBD vacuum state.

Let us recall our analogy of synchronized swimming to achieve a better understanding of the rich domain structure of the vacuum state of the water rotational field in QBD. Suppose we have a chance to see an exhibition of synchronized swimming performed by all the national teams at the Olympic Games. Imagine more than one hundred national teams about to perform their different routines simultaneously in a large swimming pool. A Canadian team, a Japanese team, an American team, a Swiss team, and so on, all stand ready to begin unique, elegant performances at the same time, as a special exhibition program for the audience. Each national team is in its lowest energy state, that is, the vacuum state, of the motion of synchronized swimming; in this state each team stands still with the swimmers' faces turned one and the same direction. However, it might happen that different national teams are in different ordered states; that is, each team is facing a different direction. For example, all the swimmers of the Canadian team stand still and turn their faces, say, toward the front entrance of the hall, while the Japanese team faces the emergency exit J, the American team faces the press box, and the Swiss team faces a big poster of Nestlé.

There we have the lowest energy state, that is, the vacuum state, of the total motion of the international exhibition team of synchronized swimmers. The audience will see that the whole team in the large Olympic pool is divided into many domains, in each of which a national group manifests order or harmony. That is the domain structure of the vacuum state of the total motion of the international exhibition team of synchronized swimming. There is no one vacuum state, but instead an infinite variety of vacuum states. In one vacuum state of the international exhibition team, all the swimmers of the Canadian team stand still and face the emergency exit K, but in another vacuum state they stand still and face the press box, and so on. There are as many varieties as the number of national teams belonging to the international exhibition team times the number of different directions the teams can face. A similar rich domain structure exists in the vacuum state of the water rotational field in QBD.

Domain structures such as QBD vacua appear frequently in the various research fields of quantum theory. Superconductivity, magnetism, and phase transitions are typical, familiar examples. For instance, in the quantum theory of magnetism, the atomic ingredients of magnetized or magnetizable matter are the magnetic-dipole-moment vectors of **atoms**. Those magnetic dipoles play the role of electric dipoles of the water rotational field, and a similar domain structure emerges in magnetized or magnetizable matter. That is, in each domain, called a magnetic domain, all the magnetic-dipole-moment vectors are aligned in one and the same direction, and they generate a constant, spontaneous magnetization. The aligned direction of the magnetic dipoles may change from domain to domain, and the total magnetization is given by a vector sum of the constant, spontaneous magnetizations of all the magnetic domains.

The Nambu-Goldstone theorem in quantum field theory well explains the domain structure of the QBD vacua (Umezawa 1993). According to this theorem, a vacuum state of a field entity governed by quantum field theory can be considered a macroscopic ordered state manifesting a spatial extension the size of the coherence length under this general condition: the quantum dynamics of the field entity is invariant under a particular symmetry transformation, but the vacuum state itself is no longer invariant; that is, it breaks the symmetry. In the case of the water rotational field, a uniform rotation of the electric-dipole-moment vectors of the H_2O molecules around their respective centers does not change the fundamental dynamics of the water rotational field described by the Schrödinger equation. The vectors remain in the same positions relative to each other; each vector rotates around its center with the same amount of directional angle as the other vectors. Such a uniform rotation does not affect the dynamics. In other words, the dynamics of the water rotational field is invariant under a rotational symmetry. However, the dynamics of the water rotational field in any QBD vacuum state breaks this fundamental symmetry, since the rotation of the electric dipole of each H_2O molecule changes the direction of the aligned electric dipole moments, thus changing one QBD vacuum state into another one. The direction of the aligned electric-dipole-moment vectors in the QBD vacuum state is not invariant and symmetrical under the rotational transformation. The vacuum state in QBD breaks the original symmetry apparent in QBD; thus, it is said to be a vacuum state of spontaneous symmetry-breaking type.

Let us get an intuitive but clear understanding of the vacuum state of spontaneous symmetry-breaking type by returning once again to our analogy. Suppose a synchronized swimming team is performing elegantly. The total movement of the team's performance does not change if the direction each teammate faces suddenly changes by the same amount of directional angle. That is, the dynamics of the synchronized swimming team is invariant under the rotational symmetry. However, this is no longer true if the team is in a vacuum state, that is, the lowest energy state, in which all the teammates stand still with their faces turned, say, toward the press box. If the directional angle each teammate faces changes to the same degree, so the group as a whole makes a uniform rotation, the team members' faces no longer point toward the press box but, say, toward the emergency exit. Thus, the team is no longer in the vacuum state characterized by facing the press box, but in another vacuum state characterized by facing the emergency exit. This means that the vacuum state, that is, the lowest energy state, of a synchronized swimming team breaks the rotational symmetry.

Any vacuum state of the water rotational field in QBD (i.e., any QBD vacuum) is indeed a vacuum state of spontaneous symmetry-breaking type, and it manifests nonlocality and the long-range order characteristic of the collective mode in quantum field theory. The QBD vacuum is a peculiar macroscopic ordered state of the atomic ingredients of matter whose occurrence cannot be explained without recourse to the action of a fundamental quantum field theoretical mechanism responsible for spontaneous symmetry-breaking. Thus, any QBD vacuum manifests a domain structure extending over about ten to the tenth domains of macroscopic ordered states, each with a spatial dimension on the order of the coherence length, that is, about 50 microns.

The Nambu-Goldstone theorem in quantum field theory asserts that the result of the spontaneous breakdown of any compact and continuous symmetry, such as the rotational symmetry, is the manifestation of massless quanta, that is, quanta with **mass** equal to zero, called Goldstone bosons. In any macroscopic ordered state, such as the QBD vacua, in which the field entity is aligned everywhere in one and the same direction, any small perturbations of the field entity's alignment appear as long-range correlation waves. Because the field entity in the ordered state is aligned in one and the same direction, a deviation of the field entity at some point in space will be propagated successively from point to point; this is regarded as a wave. The wave is maintained by the presence of macroscopic, long-range order in the macroscopic ordered vacuum state of spontaneous symmetry-breaking type, and so may be understood as a long-range correlation wave. Those long-range correlation waves in the macroscopic ordered state can be regarded also as energy quanta, according to the basic principles of quantum field theory. They are specific bosons, called Goldstone bosons, whose energy eigenvalues change continuously from zero to any positive value of energy. Recall that bosons (i.e., Bose-Einstein quanta) are energy quanta, any number of which can occupy the same state of quantum dynamics corresponding to a wave motion of the field entity.

In quantum field theory, the mass of an energy **quantum** of a field quantity is defined as the smallest amount of energy needed to create a single energy quantum, divided by the squared speed of light. In principle, an infinitesimal amount of energy is sufficient to create a Goldstone boson, so it is considered a massless Bose-Einstein quantum; that is, the mass of a Goldstone boson is zero. Any value of energy, from zero to infinity, can be the energy eigenvalue of a Goldstone boson, so Goldstone bosons with an energy eigenvalue close to zero can be created in the macroscopic domain of an ordered vacuum state by receiving any small amount of energy.

Recall that the energy of an energy quantum of a field entity divided by the

Planck constant stands for the **frequency** of the corresponding wave motion of the field entity in quantum field theory. Therefore, Goldstone bosons characteristic of the macroscopic ordered vacuum states of the QBD system can also be regarded as waves with very low frequencies. Those waves are nothing but long-range correlation waves of the water rotational field created in an ordered vacuum state, in which all the electric-dipole-moment vectors of H_2O molecules are aligned in one and the same direction. Hence, those waves can be considered as macroscopic, low-frequency, wave motions of aligned electric dipoles in the macroscopic domain of an ordered vacuum state the size of the coherence length (about 50 microns). Those low-frequency waves of aligned electric dipoles generate, of course, low-frequency **electromagnetic waves**. This may be a clue to understanding the yet unknown mechanism of the EEG (electroencephalogram), also known as the brain wave. *No microscopic physical phenomena other than the Goldstone boson can realize a macroscopic electromagnetic wave with the extremely low frequency of a brain wave.*

Goldstone bosons are massless quanta that require almost zero energy to be created, but the finite size of the macroscopic domain participating in spontaneous symmetry breakdown endows them with a small effective mass, that is, a small threshold energy for their creation. This may be understood by means of an analogy: the larger a cavity is, the lower the frequency of the sound resonating in the cavity. This frequency can be reduced even further if the size of the cavity is made larger. However, zero frequency, that is, zero energy, cannot be realized as long as the size of the cavity remains finite. We need infinite size to generate an idealistic resonating sound with zero frequency.

The Goldstone boson in QBD is nothing but a long-range correlation wave of the water rotational field created in an ordered vacuum state, in which all the electric-dipole-moment vectors of H_2O molecules are aligned in one and the same direction. It is created by any small amount of energy above the threshold value supplied by the collective modes of the dynamics of electric dipole moments along the filamentous strings of proteins. In QBD, Goldstone bosons appear as new degrees of freedom that carry the original symmetry property of the water rotational field. Given this symmetry property of Goldstone bosons, the broken symmetry in the QBD vacua can be understood to be restored by the creation of Goldstone bosons.

Section 51
The Cybernetics of QBD Vacua

There are no corticons nor photons in the QBD vacuum state; however, they are created as energy quanta and give rise to sophisticated physical phenomena through creation and annihilation dynamics if sufficient energy is supplied to the QBD system of the cortical field and the electromagnetic field. When a sufficient amount of energy is put into the QBD system from an external system—such as the system of electric dipole oscillations along the filamentous strings of proteins of the cytoskeletal structures and the extracellular matrices of the brain cells—a number of corticons and photons corresponding to the amount of energy given from the external system are created, and further creation and annihilation dynamics of corticons and photons are triggered as a chain reaction. Then, the QBD system will lose energy to the external system and will fall once again into one of the QBD vacua. The vacuum state will be maintained there until another sufficiently high amount of external energy is injected.

Let's look at this physical process from a different point of view. The cerebral cortex settled into a QBD vacuum may be regarded as a special type of automated machine, that is, a *self-reproducing* ***automaton***. This automaton interacts with the external systems by exchanging energy higher than the lowest value of energy gain sufficient to create at least one corticon or photon in the QBD system. This lowest value can be understood as a threshold value of energy exchange with the external systems required for this automaton to trigger a reaction. That is, this automaton reacts against any incoming energy (from the external systems) higher than the threshold value by creating corticons and photons in the QBD system of the cerebral cortex, and after a chain reaction of creation and annihilation dynamics of corticons and photons, it loses energy by scattering it to the external systems. Then, the automaton of the QBD system falls again into one of the QBD vacua. This automaton goes far beyond the ordinary concept of an automaton: it exhibits a self-modifying ability. In other words, the QBD system is a self-reproducing automaton.

Recall that an automaton is an apparatus that automatically performs certain actions by responding to preset conditions or encoded instructions. It acts in a mechanical way. An automaton requires a programmed set of instructions to perform a function that is fixed from the start, and the automaton cannot modify its function. We normally think of "self-modification" as a form of intelligence that allows an organism to change its own behavior. For the purpose of inventing an artificial intelligence, that is, a learning machine, an ingenious Hungarian mathematician, **Johann von Neumann**, developed the concept of a self-reproducing automaton (von Neumann 1966). It is a generalized automaton that can change its own functioning, response, or encoded instructions so that it can automatically adjust its actions optimally. Thus, the self-reproducing automaton is a theoretical model of the cybernetics invented by another gifted mathematician, Norbert Wiener.

As a matter of fact, the QBD vacuum state into which the QBD system falls after scattering energy to the external systems would be generally different from the one realized before the receipt of external energy. The reason is that there are an infinite number of QBD vacua, as shown in the preceding section. This infinite multiplicity or degeneracy of the QBD vacuum states enables the QBD system to manifest a self-modifying ability. Let us examine the process through which the QBD system, as a self-reproducing automaton, changes its dynamical state among an infinite number of vacuum states. How might we interpret this process as the action of a self-reproducing and self-modifying automaton?

Let us suppose that the QBD system in the cerebral cortex is in one of the QBD vacuum states. Then, if energy higher than the threshold value is received from an external system interacting with the QBD system, corticons and photons are created. The QBD system is excited into a higher energy eigenstate by absorbing energy from the external system. The energy in the form of corticons and photons triggers sophisticated creation and annihilation dynamics corresponding to the energy gain. Finally, the QBD system falls into one of the QBD vacua by releasing energy to the external system.

This illustrates the action of the QBD system, which can be understood as the self-modifying process of a self-reproducing automaton. How does the automaton change itself through this process? It is in a certain vacuum state before energy exceeding the threshold is supplied. This state in which the QBD system has temporarily settled could be any one of the infinitely many QBD vacua. Corticons and photons are created out of this vacuum state if external energy is supplied to the QBD system, and the creation and annihilation dynamics takes place in that vacuum. In other words, the response action of the QBD system as a self-reproducing automaton in the cerebral cortex reflects the former vacuum state: all the corticons and photons are created out of the former vacuum state. The creation and annihilation dynamics of corticons and photons terminates by scattering energy to the external systems. All the corticons and photons are annihilated, the energy goes to the external systems, and the QBD system falls again into one of its vacuum states. This new vacuum state is different from the preceding one. It results from the realignment of electric dipoles of the water rotational field in the previous vacuum state. The alignment of electric dipoles change as energy from an external system triggers the creation and annihilation dynamics of corticons and photons. That is, the QBD system in the cerebral cortex retains part of its former vacuum state during the creation and annihilation dynamics of corticons and photons induced by energy coming in from the external systems, and after the creation and annihilation dynamics, the QBD system settles into a new vacuum state as energy is released back into the external systems. The resulting QBD vacuum state is thought to reflect both the former vacuum state and the creation and annihilation dynamics triggered by incoming energy from the external systems. In other words, the QBD system equipped with a vacuum state in the cerebral cortex is a complete automaton, that is, a self-modifying machine, which changes from one vacuum state into another according to incoming energy. The automaton of the QBD system registers that there is incoming energy, and this input changes its internal state.

Let us return to our favorite image of synchronized swimming to illustrate the self-modifying process of the QBD system as intuitively as possible. Envision again an international exhibition of synchronized swimming by all the national teams at the biggest Olympic Games. There, more than one hundred national teams, including a Canadian team, an American team, a Swiss team, an Austrian team, and so on, perform simultaneously in one large swimming pool. All the national teams stand ready to begin their elegant and harmonious performances en masse as a special program for the Olympic audience. Each national team is in the lowest energy state, that is, the vacuum state, in which all the swimmers stand still and face the same direction. As we have seen before, it seems probable that the different national teams are in different ordered states, which means that they are facing different directions. Suppose that all the swimmers of the Canadian team stand still and look the same direction, say, toward the front entrance of the hall, while the American team faces the emergency exit J, the Austrian team faces the press box, and the Swiss team faces a big poster of Nestlé. Furthermore, assume that the American team is located next to the Canadian team and the Swiss team beside the Austrian team.

This is nothing but the lowest energy state, that is, the vacuum state, of the total motion of the united team of synchronized swimmers. The whole team in the Olympic pool appears to be divided into many domains in which each group of swimmers manifests order or harmony. This is the domain structure of the vacuum state of their synchronized motion, which is realized when all the swimmers stand ready to begin their elegant and harmonious team performances at the same time. This vacuum state is the initial configuration. As the music is turned on, all the national teams present their original performances of synchronized swimming, and each team ends in another vacuum state when the music is over. Thus, the united, international exhibition team in the Olympic pool modifies the vacuum state of the total motion by a performance of synchronized swimming led by the music. This is completely analogous to the QBD system modifying its vacuum state of quantum dynamics by manifesting the creation and annihilation dynamics of corticons and photons with respect to the incoming energy: All the national teams together are the QBD system, the music is the energy derived from the external systems, the routines are the dynamics, and the photons and corticons are the swimmers in motion. The swimmers still exist in the vacuum state.

Suppose that the music happens to be Stephen Foster's, and the swimmers on both the Canadian and the American teams spontaneously integrate their routines to create a single, unified performance, so the audience finds no apparent boundary between the two teams during the performance. Two adjacent teams, stimulated by the music, are unified into one team, so the final vacuum state of their motion manifests a new domain structure. The audience would not see two different national teams (Canadian and American), but a single, unified team of North America. In that case, the fact that the leading music is Foster's is stored or registered in the resulting vacuum state as a typical domain structure unifying the American and Canadian teams. Similarly, the fact that the music is a yodel is stored in the resulting vacuum state as a typical domain structure unifying the Swiss and Austrian teams. This whole analogy illustrates the essential aspect of the self-modifying process of the QBD system in response to incoming energy from the external systems: teams together are taken to be the QBD system. Furthermore, the modifying process of the QBD system can be viewed as the storing or recording of incoming energy. Once the vacuum state has been modified by the incoming energy—that is, by corticons and photons engaged in creation and annihilation dynamics—the QBD system will respond differently to energy received subsequently because the current vacuum state is no longer the same as the former one. In this sense, the QBD system is thought to be a self-reproducing and self-modifying automaton.

Chapter 9
Quantum Theory of Consciousness

Section 52
What is Consciousness?

We have found with the aid of **quantum field theory** that the QBD system in the cerebral cortex functions as a self-reproducing **automaton**. The water rotational **field**, which extends over the whole assembly of brain cells, gives rise to **energy** quanta called corticons at the sub-microscopic scale. Those corticons interact with each other by creating and annihilating **photons**, that is, energy **quanta** of the **electromagnetic field**. The quantum field theoretical dynamics of the water rotational field and the electromagnetic field is called quantum brain dynamics (QBD). The QBD system in a vacuum state reacts to energy derived from an external system if the incoming energy is higher than the threshold value. It creates corticons and photons and then falls into another vacuum state as energy is diffused back into the external system. Thus, it accomplishes a self-modifying process. The QBD system in the cerebral cortex exhibits **memory** and **consciousness** as a self-reproducing automaton described by quantum field theory.

How are the mechanisms of memory and consciousness given in QBD? What kind of fundamental processes take part there? Let us consider consciousness first and memory second. Then, the connection between memory and consciousness will be grasped within the realm of quantum field theory, and we will close our exploration into the sub-microscopic world of our cryptic universe, the brain.

What is consciousness? What is **mind**? Of course, we are not going to delve deeply into philosophical and epistemological considerations on consciousness. Rather, we are going to reveal what kind of physical phenomena might be identified with the fundamental process of consciousness from the purely physical point of view of quantum brain dynamics. The QBD system is nothing but water extending across and penetrating the whole assembly of brain cells and interacting with the electromagnetic field inside the cranium.

It may be interesting to compare the QBD system in the cerebral cortex with **Descartes**' mechanistic model of human memory and consciousness, which we covered in our discussion of the classical theory of the brain in Chapter 4. Descartes thought that brain tissue manifested a textile-like, spatial structure of interwoven nerves and that the flow of thymos created a pattern of coarsely woven textures; memory was maintained in that pattern. Consciousness, existing in the corpus pineale (the pineal gland), could observe the pattern of nerve textures by scanning the pattern of thymos streaming through them. Though the similarities are perhaps superficial, the comparison between Descartes' model and the quantum brain dynamics that we find in the sub-microscopic world of the brain is perhaps interesting. We might compare the nerves to filamentous strings of protein **molecules** and thymos to water.

We have found in quantum brain dynamics that the QBD system in a vacuum state is an automaton that reacts to incoming energy higher than the threshold value by first triggering the creation and annihilation dynamics of corticons and photons and then changing into another vacuum state as the energy is released back into the external system. Let us consider the external system that exchanges energy directly with our QBD system. Here we refer to the collective modes of the molecular vibrational field along the filamentous strings of protein molecules in the extensive, dense, three-dimensional network structure that spans the whole assembly of brain cells by patching together the cytoskeletons and the extracellular matrices. That external system directly interacting with the QBD system serves to interface between the QBD system and the macroscopic systems of dendritic and neural networks, which transmit neural impulse signals to and from bodily tissues. The external system surrounding the QBD system in the sub-microscopic world of the brain includes the microscopic system of filamentous strings of proteins and the macroscopic systems of dendritic and neural networks. That is, the external systems of the QBD system are those microscopic and macroscopic systems.

The QBD system receives incoming energy from those external systems; the energy, which might be produced by a sense organ when it receives a certain external stimulus, is transmitted first to the neural network, next to the dendritic network, and then to the filamentous strings of proteins. There is also a case in which the energy might be produced by the neural network itself, without any incoming impulses from a sense organ; in that case, it is transmitted first to the dendritic network and, secondly, to the filamentous strings of proteins. There is yet another case in which the energy might be produced by the dendritic network itself and be transmitted to the filamentous strings of proteins. Of course, the microscopic system of filamentous strings of proteins might produce the energy. In summary, the QBD system can receive incoming energy from four different sources: from the sense organ triggered by an external stimulus, from the neural network triggered by the activity of the neural network itself, from the dendritic network triggered by the activity of the dendritic network itself, and from the microscopic system of filamentous strings of proteins triggered by its own activity. Thus, the QBD system is activated by incoming energy from external systems even if no external stimuli are received by the sense organs. (Presumably dreams are created in this fashion.)

The QBD system at the sub-microscopic scale always interacts with the microscopic system of filamentous strings of proteins by exchanging energy. That exchange energy between the QBD system and the microscopic system of filamentous strings of proteins may be high enough to create a great quantity of corticons and photons in the QBD system. As long as energy is flowing into the QBD system from the microscopic system of filamentous strings of proteins, the QBD system maintains large scale, sophisticated creation and annihilation dynamics of corticons and photons. Of course, some of those corticons and photons might be annihilated after their energy is passed to the microscopic system of filamentous strings of protein molecules; that energy is relayed in the form of the collective mode of a coherent **wave** of the molecular vibrational fields. The collective dynamics of corticons and photons, created in the QBD system through the interaction with the microscopic system of filamentous strings of proteins, will indirectly affect the macroscopic dynamics of dendritic and neural networks by directly affecting the microscopic system of filamentous strings of proteins. In this way, corticons and photons can be thought to control the dynamics of the macroscopic systems of dendritic and neural networks as a result of their creation and annihilation dynamics in the QBD system.

Therefore, the dynamics of corticons and photons may generate neural impulses in the neural networks of the cerebral cortex through the microscopic system of filamentous strings of protein molecules. The neural impulses transmitted along the nerves, in turn, give rise to certain biological changes in various bodily tissues. In other words, the creation and annihilation dynamics of corticons and photons in the QBD system is capable of controlling the microscopic system of filamentous strings of proteins, the macroscopic systems of dendritic and neural networks, and, ultimately, specific functions of the human body, such as speech and motion.

Consciousness or mind is the entity that continually grasps external stimuli (i.e., physical actions of the external world on the sense organs of the human body) and controls various bodily tissues by referring to the organism's total memory, that is, the memory of all one's experiences. *With the help of quantum field theory, we have found the creation and annihilation dynamics of corticons and photons in the QBD system in the sub-microscopic world of the brain to be the entity we call consciousness or mind.* Therefore, the human mind or consciousness can be considered the totality of quantum physical processes of indefinite numbers of corticons and photons that are created and annihilated in the cryptic universe of the brain itself.

If incoming energy from the microscopic system of filamentous strings of proteins could not create corticons and photons in the QBD system, the creation rate would be much smaller than the annihilation rate. Consequently, there would be no corticons nor photons. In other words, the QBD system would be maintained in one of the QBD vacua until corticons and photons were created by newly incoming energy. In quantum brain dynamics, which regards consciousness as the creation and annihilation dynamics of corticons and photons of the QBD system, it would be natural to assume that there exists no consciousness in such a situation. The QBD vacuum state would be a state literally without consciousness. It might be interesting to investigate meditation in this respect. During meditation, a monk shuts out idle thoughts from his mind. Since mind is nothing but the action of consciousness, eliminating idle thoughts from his mind would correspond to cutting off the incoming energy transmitted from the dendritic and neural networks. Put another way, consciousness, that is, the creation and annihilation dynamics of corticons and photons of the QBD system, would be subject only to the incoming energy originating from the microscopic system of filamentous strings of proteins. The result would be a cessation of mental activity that manifests as thought.

Section 53
Memory and the Vacuum State

External actions received by various bodily tissues, especially the sense organs, are transmitted to the macroscopic system of neural networks in the cerebral cortex in the form of neural impulse signals through the peripheral nervous system. Then, two dynamics—the macroscopic dynamics of neural networks in the form of transmembrane ionic diffusions and the dynamics of charged elements in the perimembranous regions of brain cells in the dendritic networks—affect the microscopic dynamics of coherent wave propagations or collective modes of molecular vibrational fields along the filamentous strings of protein molecules. As a result, the QBD system receives energy from the adjacent microscopic system of filamentous strings of proteins, and newly created corticons and photons engage in the dynamics of existing corticons and photons to produce new quantum physical processes. The creation and annihilation dynamics of corticons and photons in the QBD system is thought to be consciousness in quantum brain dynamics.

There is much speculations about the true nature of consciousness in philosophy. We would add that consciousness should be grasped in connection with total memory, that is, the whole of memory ever stored from the very beginning of the activity of the brain in question. Our intuition is that consciousness exists on that total memory and enlarges itself by referring to that total memory. Then where is memory found within the framework of quantum brain dynamics if consciousness is the creation and annihilation dynamics of corticons and photons in the QBD system of the cerebral cortex?

As we have already seen, the peculiar features of memory, specifically its stability and nonlocality, attracted Umezawa's attention. He developed an interest in consciousness as a result of that observation. Thereafter, he looked for physical phenomena in the sub-microscopic world of the brain, which quantum field theory might explain as the fundamental process of a brain realizing consciousness. Indeed, he was inspired with the idea that memory could be the vacuum state.

The QBD system in the sub-microscopic world of the brain has been shown to manifest a self-modifying ability that relies on the specific characteristics of the QBD vacuum states; therefore, the QBD system can be regarded as a self-reproducing automaton. There, due to the interaction between the water rotational field and the electromagnetic field as well as the symmetry property of the former field, an infinite number of vacuum states can be realized as approximate minimum energy eigenstates. Suppose that the QBD system is in one of those QBD vacua. That means the water rotational field extending across the whole assembly of brain cells in the cerebral cortex manifests an overall domain structure. That structure is composed of about ten to the tenth power macroscopic domains whose scales are approximately 50 microns. The electric-dipole-moment **vectors** of all the H_2O molecules in each domain are aligned in one and the same direction up to random variations corresponding to quantum fluctuations. In a sense, a nonlocal order exists in the vacuum state of the QBD system. All the **electric dipole moments** are aligned in one and the same direction in each domain, thus manifesting an order that extends throughout that domain.

If energy is provided by the macroscopic system of neural and dendritic networks through the microscopic system of filamentous strings of protein molecules, then corticons and photons are created out of this vacuum state, and they manifest physical processes that depend on both the vacuum state and the incoming energy. Those physical processes can be well illustrated by the creation and annihilation of many corticons and photons. As a result of those physical processes, which are triggered by incoming energy, the vacuum state of the QBD system is transformed into a new vacuum state. The familiar term *vacuum phase transition* in quantum field theory refers the phenomenon in which a vacuum state changes subject to a certain physical process. For example, in quantum cosmology, the vacuum phase transition has proven indispensable to our understanding of the sudden expansion of our universe at the very beginning of its creation.

In the QBD system, a vacuum phase transition is induced by incoming energy from the microscopic system of filamentous strings of protein molecules. *Seen from a different point of view, the phenomenon of vacuum phase transition may be understood to be a self-modifying process in which incoming energy changes the QBD system and the result of modification is maintained as the transformed vacuum state. In this sense, the QBD system realizes a "learning" ability.* The result of "learning," that is, self-modification in response to incoming energy, is maintained as a new vacuum state. The transformation is accomplished through the process of vacuum phase transition, so the total memory is the vacuum state itself in QBD, just as Umezawa anticipated about thirty years ago.

In search of the memory trace in the brain, that is, the *engram*, we found our way into the sub-microscopic world of our cryptic universe in the brain. There we discovered the vacuum state of the QBD system fully described by quantum field theory. *The engram is revealed to be the domain structure of a QBD vacuum state, in which the electric-dipole-moment vectors of all the H_2O molecules, aligned in one and the same direction, manifest macroscopic order in each domain up to quantum fluctuations.*

We regard the creation and annihilation dynamics of corticons and photons as the physical process of consciousness, which can be created out of the existing vacuum state by incoming energy from the macroscopic system of neural and dendritic networks through the microscopic system of filamentous strings of protein molecules. Therefore, consciousness might emerge from the existent total memory by receiving energy from those external systems. Consciousness, seen as a physical process among corticons and photons triggered by external stimuli, might induce a vacuum phase transition and change one vacuum state into another one, thereby enlarging total memory itself. Thus, the vacuum state necessarily turns out to be total memory as long as the creation and annihilation dynamics of corticons and photons are regarded as consciousness in quantum brain dynamics. *Simply put, consciousness-memory-consciousness-memory-consciousness is a continuous process. The two "feed" each other, so to speak.*

The principles of quantum brain dynamics that originated from the works of Umezawa, Stuart, and Takahashi have enabled us to discover the trace of memory, that is, the *engram,* in the sub-microscopic world of the brain. It is nothing but the stable and nonlocal domain structure of the QBD vacuum state in which the electric-dipole-moment vectors of all the H_2O molecules are aligned in one and the same direction in each macroscopic domain up to quantum fluctuations. Recall the image of the domain structure of the vacuum state of synchronized swimming when all the swimmers from numerous teams stand ready to start their elegant and harmonious performances simultaneously. Such a stable and nonlocal manifestation of the QBD vacuum state is realized essentially by the symmetry property of the water rotational field extended to the whole assembly of brain cells and its interaction with the electromagnetic field. These findings might put an end now to a long-standing question about memory and consciousness: *The mechanisms of memory and consciousness do manifest nonlocality, just as Lashley and Pribram hypothesized.* This is the logical conclusion of Umezawa's quantum field theory of the brain, that is, quantum brain dynamics. Umezawa might say, *"The essential aspect of the vacuum state of interacting fields in quantum field theory is nonlocality manifesting a long-range order, my friends."*

Section 54
Quantum Cosmology of the Brain—Part One

A great many of the physical phenomena unfolding in our universe can be understood in a unified way in terms of the fundamental principles of physics in the sub-microscopic world, that is, **quantum mechanics** and quantum field theory. Modern high energy physics seems to be full to the brim with the greatest attractions ever found: quantum electrodynamics, quantum chromodynamics, quantum cosmology, and grand unified theories that include those others.

Similar dynamics might be at work in our cryptic universe, the brain. Umezawa and his collaborators, Takahashi and Stuart, made it clear that quantum field theory plays an important role in the sub-microscopic world of the brain and that the fundamental processes of the brain that give rise to memory and consciousness might be grasped within the realm of quantum brain dynamics. A completely new scientific way of understanding much of the mental activity unfolding in our cryptic universe, the brain, has been made possible through a quantum theoretical approach. Let us now consider the fundamental processes of the brain, as revealed by quantum brain dynamics, to develop a quantum cosmology of the brain.

We have already seen that corticons and photons, quanta of the water rotational and electromagnetic fields, are created and annihilated in the sub-microscopic world of the brain, that is, within the QBD system. Corticons and photons are created in the QBD system by incoming energy from the macroscopic system of neural and dendritic networks through the microscopic system of filamentous strings of proteins. The incoming energy triggers the creation and annihilation dynamics of corticons and photons. This dynamical event is regarded as the physical process of consciousness. Then, are corticons and photons the only quanta that color the cryptic universe of the brain? The answer might be negative. As a matter of fact, there are two other types of quanta present and active in the cryptic universe of the brain. A quantum cosmology of the brain might not be complete if those quanta are overlooked.

One of those quanta is that which appears in the physical process of memory recall. It is called a *Goldstone boson*. The other is involved in memory loss. It is called an *instanton*.

There is some evidence that our understanding of consciousness as reliant upon total memory is correct. Consciousness is revealed to be the creation and annihilation dynamics of corticons and photons; the corticons and photons are created by incoming energy. That energy from the macroscopic system of neural networks first passes through the microscopic system of filamentous strings of proteins and then triggers the creation and annihilation dynamics of the QBD system out of the vacuum state representing the total memory stored since the creation of the brain and up to the moment of triggering. According to the fundamental principles of quantum field theory, energy higher than a certain threshold value is required to create corticons and photons in the QBD system. In this sense, energy that can create at least one corticon or photon must be supplied to perpetuate consciousness in the microscopic world of the brain; that is, the creation and annihilation dynamics of corticons and photons, sustained by incoming energy, constitutes consciousness. Put another way, one's consciousness reacts to the external stimulus through the microscopic system of filamentous strings of proteins, and if a vacuum phase transition is induced, then the external stimulus together with the existing total memory realizes another vacuum state as a new total memory and, consequently, new consciousness. It might be helpful to think of total memory as the continuous unfolding of one state into another in the ongoing emergence of consciousness, like a trunk of a tree that makes a growth ring, retaining its past form while developing into a new form.

Within the realm of quantum brain dynamics, how can consciousness, which is regarded as the totality of physical processes engaged in the creation and annihilation of corticons and photons, recall memories? The vacuum phase transitions are induced frequently by energy from external systems. Therefore, this question naturally arises: What is the mechanism by which the dynamics of corticons and photons peruse the existing memory? Umezawa's favorite strategy in quantum field theory comes into play once again.

Here, as in preceding sections, we encounter the peculiar case of a field entity that manifests a certain symmetry property, but not in the vacuum state. The vacuum state in the quantum brain dynamics examined here might be an example of such a peculiar phenomenon. The electric-dipole-moment vector of each H_2O molecule of the water rotational field has a symmetry property. The quantum dynamics of the water rotational field can remain invariant, that is, unchanged, even as the electric-dipole-moment vectors of all the H_2O molecules change directions by the same spatial angle. In other words, the water rotational field in QBD manifests a rotational symmetry. However, in a vacuum state, the electric-dipole-moment vectors of all the H_2O molecules are aligned in one and the same direction; they manifest a long-range order characteristic of collective modes in quantum field theory. In such a situation, the original symmetry property of the water rotational field in QBD is broken; the vacuum state in which the electric-dipole-moment vectors of all the H_2O molecules are aligned in one particular direction cannot remain invariant as the electric-dipole-moment vectors of all the H_2O molecules change directions by the same spatial angle. That is, one vacuum state changes to another vacuum state. This is a typical example of the quantum field theoretical notion of spontaneous symmetry-breaking.

As shown by the Nambu-Goldstone theorem in quantum field theory, a vacuum state of the spontaneous symmetry-breaking type is known to create massless quanta, called Goldstone bosons, with any small amount of energy. Recall the following: In any macroscopic ordered state, such as the QBD vacuum, in which the field entity is aligned everywhere in one and the same direction, any small perturbations of the field entity's alignment appear as long-range correlation waves. Because the field entity is in the ordered state, with all of its **degrees of freedom** aligned in the same direction, a deviation of the field entity at some point in space will be propagated successively from point to point and regarded as a wave. The wave is maintained by the presence of macroscopic, long-range order in a macroscopic ordered vacuum state of the spontaneous symmetry-breaking type; therefore, it is considered a long-range correlation wave. In quantum field theory, the **mass** of an energy **quantum** of a field entity is defined as the smallest amount of energy needed to create a single energy quantum divided by the speed of light squared. In principle, even an infinitesimal amount of energy is sufficient to create a Goldstone boson, so it is considered a massless Bose quantum; that is, the mass of a Goldstone boson is zero. This is so because any value of energy from zero to infinity can be the energy eigenvalue of a Goldstone boson, so Goldstone bosons with energy eigenvalues close to zero can be created in the macroscopic domain of an ordered vacuum state by receiving any small amount of energy.

Therefore, in the vacuum state regarded as the totality of memory traces, that is, *engrams*, in quantum brain dynamics, Goldstone bosons can be created by a very small amount of energy and can participate in the quantum field theoretical dynamics of the QBD system. It might not be too difficult to grasp the concept of Goldstone boson intuitively as follows:

A vacuum state in quantum brain dynamics manifests a macroscopic domain structure in which the electric-dipole-moment vectors of all the H_2O molecules in each domain are in a macroscopic ordered state; that is, they are aligned in one and the same direction up to quantum fluctuations. All the vectors generally point in one and the same direction, but with slight fluctuations in that the direction occasionally changes to a small degree. Suppose that a very small amount of energy is passed from the interfacing microscopic system of filamentous strings of proteins. If it is much smaller than the threshold value of energy required to create at least one corticon or photon, there will be no creation and annihilation dynamics in the QBD system of the cerebral cortex; there will be no vacuum phase transition. However, in the vacuum state, in which the original symmetry property of quantum brain dynamics is violated, a wave phenomenon will take place.

Reacting to any small amount of energy given by the microscopic system, the electric-dipole-moment vector of a certain H_2O molecule in one of the domains of a macroscopic ordered state may be slightly disoriented. An infinitesimal amount of energy suffices to change the direction in which the vector points. Recall that the electric-dipole-moment vectors of all the H_2O molecules of the water rotational field in this domain are macroscopically ordered, so they retain their alignment in that same direction. Therefore, once a perturbation is added to the direction of the electric dipole moment of an H_2O molecule, the vacuum state of the QBD system of the cerebral cortex acts to restore the alignment of the disoriented dipole moment, so that the macroscopic order in the domain of the collective mode will be maintained. Restated more simply, because there is strong nonlocal correlation among the electric dipole moments in the domain to align them in one and the same direction, the disturbing motion of the electric dipole moment propagates to neighboring ones, thus manifesting a wave motion. This is a wave phenomenon in which a small disturbance propagates in the field of nonlocal order of the QBD vacuum state. A Goldstone boson is precisely that wave motion described within the framework of quantum field theory.

In quantum brain dynamics, Goldstone bosons are created even if only a small amount of energy is transmitted from the microscopic system of filamentous strings of protein molecules to the QBD system, for total memory is stored as a vacuum state of the spontaneous symmetry-breaking type. (Here it is important to recall that the most recent alignment of the vectors is the whole memory of incoming energy ever given to the QBD system by external systems.) Those Goldstone bosons take part in the sub-microscopic physical processes in the QBD system of the cerebral cortex. Then, if consciousness exists, that is, if corticons and photons are being created and annihilated in the QBD system, Goldstone bosons characteristic of the ordered domains of the vacuum state—that is, the memory traces or *engrams*—participate in realizing consciousness; they are created through a continual sequence of infinitesimal amounts of incoming energy. In that sense, consciousness is thought to recall memory in terms of Goldstone bosons; a memory trace as an ordered domain of the QBD vacuum can be recalled by creating Goldstone bosons within the QBD system. *The creation and annihilation dynamics of corticons and photons interacting with those Goldstone bosons is the fundamental process of consciousness recalling memory.*

Let us illustrate this fundamental process of recalling memory with the help of our favorite image, synchronized swimming. Envision again our exhibition of synchronized swimming by all the national teams at the Olympic Games. More than one hundred national teams—including an American team, a Canadian team, a Swiss team, an Austrian team, and so on—perform simultaneously in a huge swimming pool. Let us return to the point at which each national team is in the lowest energy state, that is, the vacuum state, of the motion of synchronized swimming: the swimmers in each group stand still and face the same direction. For example, all the swimmers of the American team face the front entrance of the hall, while the Canadian team points toward the emergency exit J, the Austrian team toward the press box, and the Swiss team toward a big poster of Nestlé. Furthermore, recall that the American team is located beside the Canadian team and the Swiss team by the Austrian team.

A single cohesive team, composed of all the national teams, appears to be divided into many domains in which the swimmers manifest order or harmony. This is the domain structure of the vacuum state of synchronized swimming, which occurs when all the swimmers are standing ready to start their performances at the same time. This vacuum state is the initial configuration of the total motion of the synchronized swimming exhibition. As the music is turned on, all the national teams perform their original synchronized swimming routines. They end in another vacuum state of the international exhibition team when the music is over.

Recall that Stephen Foster's music serves to unify the motion of the Canadian and American swimmers into one spontaneous, synchronized routine, so the audience finds no apparent boundary between those two teams during the performance. The two adjacent teams spontaneously blend into one team as they move to the same music, and the final vacuum state in which the motion settles manifests a new domain structure. Then the audience no longer sees the Canadian and American teams, but a single, large unified team of North America. In that case, Foster's music is stored or registered in the resulting vacuum state as a typical domain structure that unifies the American and Canadian teams. Similarly, if the accompanying music is a yodel, that is stored in the resulting vacuum state as a typical domain structure that unifies the Swiss and Austrian teams. Meanwhile, the other domains remain unchanged.

For the moment, let us focus on the ordered domain of the North American team in which all the swimmers stand still and turn their faces the same direction, say, toward the press box this time. This new ordered domain is nothing but the memory trace of the previous musical accompaniment, namely, Foster's. Now suppose that someone in the audience close to the front entrance shouts suddenly. Instantly one of the swimmers in the ordered domain of the North American team turns her face toward the shout, that is, toward the front entrance, and then returns to the team's original direction of alignment. The fact that all the swimmers of the North American team are in the ordered state implies that each swimmer is sensitive to the motion of a neighboring teammate and immediately follows the motion. In other words, each element of the field entity in a collective mode manifesting macroscopic, long-range order mirrors the behavior of a neighboring element. That is, a teammate next to the perturbed swimmer is sensitive to the disruption and immediately follows suit by turning her face toward the front entrance. Through a chain reaction, or a kind of domino effect, after a brief glance at the front entrance, the swimmers all realign themselves to face the press box again, so the original alignment is restored.

In that way, the perturbation of one swimmer's alignment propagates from teammate to teammate successively throughout the ordered domain of the North American synchronized swimming team in its vacuum state. This domino effect, so to speak, is seen as a wave by the audience. This wave is characteristic of the ordered domain of the North American team. The Goldstone boson can be well illustrated by that kind of wave propagation in the ordered domain of the vacuum state.

Section 55
Quantum Cosmology of the Brain—Part Two

Finally, let us look at another type of quantum that plays a part in the fundamental processes of the brain: the instanton. We have already revealed that *total memory is nothing but one of the QBD vacuum states and that consciousness is the creation and annihilation dynamics of corticons and photons in the QBD system of the cerebral cortex.* If the energy that comes from the macroscopic system of neural and dendritic networks through the microscopic system of filamentous strings of proteins exceeds a certain threshold value, then new corticons and photons are created in the QBD system, and their dynamics alter consciousness. The vacuum state of the totality of memory existing at that time undergoes a vacuum phase transition, and another vacuum state is realized as a new total memory. The new vacuum state reflects both the pre-existent total memory and the change of consciousness triggered by the incoming energy.

Quantum brain dynamics suggests that total memory is continually altered by consciousness, triggered by energy exceeding the threshold value required to create at least one corticon and photon in the QBD system received from the macroscopic system of neural and dendritic networks through the microscopic system of filamentous strings of proteins. *The continual change of the total memory is thought to be controlled by consciousness, and it is nothing but an integration of all the modifications that the QBD system of the cerebral cortex has made successively as a self-reproducing automaton.*

In such a mechanism of memory in quantum brain dynamics, however, total memory is always perfectly retained. From the moment of birth, it stores the entire sequence of external stimuli received as physical actions—that is, the sum of a person's interactions with the external world—as well as the entire sequence of energies created by the macroscopic systems of neural and dendritic networks and the microscopic system of filamentous strings of proteins in the brain, including all that one has thought, considered, or imagined. Since one's consciousness is always capable of accessing perfect memory, that is, the total recall of that entire sequence of events, in terms of Goldstone bosons, the brain functions as a perfect self-reproducing automaton. There is no chance of the obliteration of memory.

Nevertheless, everyone has experienced memory becoming less complete as time passes, and part of it finally vanishing. Of course, we might think that total memory is always perfectly maintained in some form but that it is simply not in our awareness or that it is recalled incompletely. Let us investigate the stability of actual memory from the point of view of quantum brain dynamics.

We have clarified that one's total memory is nothing but one of the vacuum states of the QBD system. In quantum brain dynamics, the vacuum states are not true minimum energy states but approximate ones. They manifest domain structures of the water rotational field in which the electric-dipole-moment vectors of all the H_2O molecules are aligned in one and the same direction in each domain up to quantum fluctuations. In such a vacuum state, there exists a nonlocal order, that is, a quantum theoretical, long-range correlation that spontaneously breaks the symmetry property of quantum brain dynamics. Recall that the water rotational field originally has a specific symmetry property in that the quantum dynamics of the water rotational field can remain invariant, that is, unchanged, even if the electric-dipole-moment vectors of all the H_2O molecules rotate and orient themselves in a new direction. Corresponding to the number of choices of spatial angle, there are, indeed, an infinite number of vacuum states. According to the principles of quantum field theory, among those infinitely many vacua, only one vacuum state is realized as the total memory in the QBD system. The total memory, that is, the existing vacuum state, will suffer a vacuum phase transition if energy higher than the threshold value is supplied by the macroscopic system of neural and dendritic networks through the microscopic system of filamentous strings of proteins. With the minimum amount of energy to create at least one corticon or photon in the QBD system, the existing vacuum state will be transformed into another one. The total memory is subject to such evolution.

Then, does the evolution of memory, that is, the evolving QBD vacuum state, arise only from the vacuum phase transition induced by the sub-microscopic physical process of creating and annihilating corticons and photons as a result of incoming energy from the external systems? The answer is no. There is a mysterious phenomenon peculiar to **quantum theory**. It is called the **quantum tunneling** phenomenon. Quantum tunneling occurs because the QBD vacuum state is an approximate vacuum state and not a true minimum energy eigenstate.

Quantum tunneling phenomena are extraordinary physical phenomena in the sub-microscopic world, which can be understood only within the theoretical framework of quantum mechanics and quantum field theory. Suppose an **electron** is trapped in a sub-microscopic pitfall. As shown in Chapter 2, the motion of a quantum, such as an electron, can only be portrayed by the Erwin meter, that is, the wavefunction in terms of **probability**. The fact that an electron is trapped in a sub-microscopic pitfall is expressed in quantum mechanics as follows: the probability of finding an electron, given by the squared length of the hand of the Erwin meter, namely the **absolute square** of the wavefunction, is very high in the pitfall and very low outside. In terms of energy, we may also say that an electron is trapped within a sub-microscopic pitfall when the energy of that electron is less than the value needed to get over the wall of the pitfall. This is true in **classical mechanics**; no material body can escape the pitfall if its energy is less than the value needed to climb over the wall. However, in quantum mechanics, there is a strange phenomenon in which an electron trapped in a sub-microscopic pitfall does get out of the pitfall, even if its energy is less than the requisite value.

Solving the Schrödinger equation governing the wavefunction of an electron in such a predicament reveals that the value of the wavefunction does not become zero in the spatial region outside the pitfall. The probability of finding an electron outside the sub-microscopic pitfall is non-zero. In quantum mechanics, any physical event that has a non-zero probability of occurrence can actually occur. Therefore, in quantum mechanics, an electron with energy less than the value needed to get over the wall of the pitfall can surmount this obstacle because the probability of finding the electron outside the pitfall is non-zero. This extraordinary phenomenon, which is typical in quantum theory, is called a quantum tunneling phenomenon; the electron escapes the pitfall as if by making a tunnel through the wall.

If the QBD vacuum state were an exact minimum energy eigenstate of the QBD system, then it would satisfy the stationary Schrödinger equation for the stationary wavefunction of the water rotational field; it would remain constant in time and manifest a completely stable state. However, the QBD vacuum state is only an approximate energy eigenstate and might not satisfy the stationary Schrödinger equation in question. The QBD system of the cerebral cortex in such an approximate vacuum state is no longer stable; it is rendered unstable. Since the time evolution of the QBD vacuum state is given by the Schrödinger equation, we can estimate how such an approximate vacuum state changes in time. As a result, we find that the probability that the QBD vacuum state will be transformed into another one increases as time passes, even if no energy enters the QBD system from the external systems. In quantum mechanics and quantum field theory, nothing can be prescribed without using probability. This fact—that there is increasing probability of the QBD vacuum state transforming into another one—means that eventually it will change into another vacuum state, thus realizing a spontaneous vacuum phase transition.

The vacuum state, that is, total memory in quantum brain dynamics, might suffer a spontaneous vacuum phase transition due to the quantum tunneling effect. That means it might change into another vacuum state without an apparent cause. Furthermore, the resulting vacuum state would be chosen independently of the actual creation and annihilation dynamics of corticons and photons, due to the purely probabilistic nature of the quantum tunneling phenomenon. In other words, a distortion of total memory might be produced in the QBD system, a distortion that cannot be controlled by consciousness. In quantum brain dynamics, the phenomenon of spontaneous vacuum phase transition due to the capricious quantum tunneling effect makes memory less complete. A vacuum phase transition due to a quantum tunneling phenomenon is not infrequent in quantum field theory. For example, in modern quantum cosmology, the possibility exists that a universe created in one vacuum state might change into another universe created in another vacuum state, due to the quantum tunneling effect.

Spontaneous vacuum phase transition due to the quantum tunneling phenomenon does not necessarily take place uniformly in each domain of the collective mode. For example, in the phase transition of water into vapor due to incoming thermal energy, only some parts of the water, not *every* part of the water, suffers the phase transition; thus, many bubbles appear uncontrollably. In a similar manner, the spontaneous vacuum phase transition in the QBD system might progress in such a way that random bubbles of other QBD vacua appear uncontrollably everywhere in the domain structure of the macroscopic ordered state of the QBD vacuum. In quantum field theory, such bubbles of vacua can be thought of as quanta, and they are called *instantons*. Therefore, the vacuum state standing for total memory in quantum brain dynamics is not a completely stable entity. It suffers from the creation of an indefinite number of instantons as time passes. The number of instantons increases over time.

Instantons emerge in the QBD vacuum state representing total memory as bubbles of other QBD vacuum states created by the spontaneous vacuum phase transition when the capricious and mysterious quantum tunneling effect is only partially successful. There will be an increasing number of holes in the total memory made by other vacuum states if we let the instantons take their own course. To preserve memory from the decay that results from the emergence of instantons, we have to continually induce vacuum phase transitions controlled by consciousness. That is, the creation and annihilation dynamics of corticons and photons sustain the continual evolution of our consciousness, so there is little chance for instantons to interrupt. The old teaching that the less you think, the weaker your brain might be well supported by quantum brain dynamics. A human being was called a "thinking reed" in ancient Egypt. Quantum brain dynamics reveals that he or she is also a quantum theoretical, self-reproducing automaton that cannot stop thinking.

Epilogue

In this volume, the authors have presented a theoretical framework, called quantum brain dynamics, to investigate consciousness scientifically in light of the first principles of physics, that is, the most fundamental laws of quantum field theory. This framework is based on the original physical theory of memory and brain functioning found in the quantum field theory developed by Ricciardi and Umezawa in the 1960s and by Stuart, Takahashi, and Umezawa in the 1970s. Although not as popular as quantum mechanics for brain and life scientists, quantum field theory is, indeed, a fundamental theory in modern physics. All the physical theories, including quantum mechanics, can be derived from quantum field theory. It is in this sense that quantum field theory is thought to provide the first principles of physics. In recent years, much attention has been paid to the unified theory of elementary particles, the grand unified theory of everything including gravitation, and quantum cosmology, which reveals the creation and annihilation of our universe as well as others; however, little notice has been paid to the fact that these theories were all developed within the realm of quantum field theory.

It is true that several different authors have attempted to develop a scientific framework for investigating consciousness in terms of quantum mechanics. Among them are Bohm (1951), Bohr (1958), Eccles (1986), and Penrose (1989; 1994). Their common and principal interest has been in the acausal and nonlocal character of the wavefunction frequently discussed in the problem of quantum measurement: The wavefunction of a sub-microscopic system of matter suffers from the observer's consciousness in an acausal and nonlocal way; thus, interesting philosophical interpretations of the relationship between the wavefunction of quantum mechanics and consciousness were proposed. It must be emphasized, however, that quantum mechanics is a theoretical framework of physics that can be applied safely to any sub-microscopic system of matter composed of a small number of elementary particles isolated from other systems of matter in the universe. Therefore, it might be a reckless attempt to apply the specific framework of quantum mechanics to a macroscopic system of matter such as the brain or the living cell made of ten to the tenth power of elementary particles. Those macroscopic systems of matter are certainly beyond the limit of direct application of quantum mechanics, even though the problem of quantum measurement seems quite interesting from a philosophical point of view.

Due to simplistic applications of quantum mechanics to the investigation of brain functioning and consciousness, several physicists working in the non-basic domains of research sounded an alarm against the incorporation of any quantum theoretical idea into both brain and life sciences. As a result, many brain scientists have developed a false idea that all the quantum theoretical approaches to the fundamentals of brain functioning and consciousness are equally nonsensical, without recourse to a correct understanding of quantum theory. Indeed, they believe that classical physics might be the proper realm of both brain and life sciences. It must be mentioned explicitly, however, that classical physics is no more than a phenomenological framework of physics adequate for describing the dynamics of macroscopic matter.

The first two basic research questions—"What is reality?" and "What is the universe?"—have been solved by discovering the first principles of fundamental physics, quantum mechanics, and quantum field theory. Therefore, from the unified point of view of those basic researches, even macroscopic matter would fall within the scope of quantum mechanics and quantum field theory, as long as macroscopic matter is a part of the reality of our universe. This means that classical physics could be derived from the first principles of fundamental physics. Although many physicists have recklessly believed that classical physics can be obtained from quantum physics by simply passing the Planck constant to zero, a deep consideration by Bohr (Umezawa 1993) prevented us from such an easy way of realizing the classical physics of macroscopic matter. Indeed, only recently has the theoretical framework of quantum field theory disclosed that macroscopic matter is realized as a particular quantum theoretical state of the collective dynamics of sub-microscopic matter with a macroscopic manifestation of order (Umezawa 1993). This interesting fact seems to force us to consider the last two basic researches from the unified point of view of the first two basic researches; that is, we have to trigger a quantum revolution of brain and life sciences. This time, no one can claim that classical physics is sufficient to investigate macroscopic matter such as the brain or the living cell because both the appearance of macroscopic matter and the validity of the phenomenological framework of classical physics are justified only by working with quantum field theory. The readers are cordially invited to read the exposition of a modern unified theory of physics derived from quantum field theory by Umezawa in Appendix A of this volume.

What on earth is the quantum revolution of brain and life sciences? It is, indeed, the aim of this volume to provide brain scientists with both the pedagogic instruction in quantum mechanics and quantum field theory needed to understand the quantum revolution and a detailed story of the quantum revolution of brain science. The quantum revolution of life science has been addressed only briefly in Sections 46 and 47. We shall return to this fascinating story in another volume.

The quantum revolution during the early part of this century has become convention in physics, but, as we have shown, in recent decades a new frontier of that revolution has opened up with Umezawa's proposal that quantum theoretical degrees of freedom be admitted to brain science. Perhaps this will permit, at long last, an explanation of consciousness itself, which has hitherto fallen outside the nomological net of science. We have attempted to further the development of quantum brain dynamics and its relation to consciousness in the preceding chapters.

It must be admitted, however, that the quantum revolution in brain science remains a threatened revolution. Much work is yet to be done, for example, in experimental tests of the theory, in the unraveling of philosophical perplexities pertaining to the relationship between consciousness and neural matter with quantum theoretical degrees of freedom, and in a rapprochement with cognitive science, the burgeoning science of mind. In any case, we hope that we have contributed to the magnificent quantum cosmology of brain functioning and enriched your enjoyment of the endless world of your brain and your consciousness.

References

Bartlett, F. and John, E. R. 1973. Equipotentiality quantified: the anatomical distribution of the engram. *Science 181*, 764-767.

Bohm, D. 1951. *Quantum Theory*. New York: Prentice-Hall.

Bohm, D. 1952. A suggested interpretation of the quantum theory in terms of "hidden variables," I and II. *Physical Review 85*, 166-179 and 180-193.

Bohm, D. 1980. *Wholeness and the Implicate Order*. London: Routledge & Kegan Paul.

Bohm, D. 1986. A new theory of the relationship of mind and matter. *Journal of the American Society for Psychical Research 80*, 113-135.

Bohr, N. 1958. *Atomic Physics and Human Knowledge*. New York: John Wiley & Sons.

Davydov, A. S. 1979. Solitons in molecular systems. *Physica Scripta 20*, 387-394.

Eccles, J. C. 1986. Do mental events cause neural events analogously to the probability fields of quantum mechanics? *Proceedings of the Royal Society of London B277*, 411-428.

Franks, F. 1972. *Water: A Comprehensive Treatise*. New York: Plenum.

Fröhlich, H. 1968. Long-range coherence and energy storage in biological systems. *International Journal of Quantum Chemistry 2*, 641-649.

Fröhlich, H. 1970. Long range coherence and the actions of enzymes. *Nature 228*, 1093.

Genberg, L., Richard, L., McLendon, G. and Dwayne-Miller, R. V. 1991. Direct observation of global protein motion in hemoglobin and myoglobin on picosecond time scales. *Science 251*, 1051-1054.

Genzel, L., Kremer, F., Poglitsch, A. and Bechtold, G. 1976. Relaxation processes on a picosecond time scale in hemoglobin and poly observed by millimeter-wave spectroscopy. *Biopolymers 22*, 1715-1729.

Globus, G. 1995. *Postmodern Brain.* Amsterdam: John Benjamins.

Gray, C. M. and Singer, W. 1989. Stimulus-specific neuronal oscillations in orientation of cat visual cortex. *Proceedings of the National Academy of Science 86*, 1698-1702.

Hameroff, S. R. 1987. *Ultimate Computing: Biomolecular Consciousness and NanoTechnology.* Amsterdam: North-Holland.

Hiley, B. J. and David Peat, F. 1987. *Quantum Implications: Essays in honour of David Bohm.* London: Routledge & Kegan Paul.

Jibu, M. and Yasue, K. 1992. A physical picture of Umezawa's quantum brain dynamics. In *Cybernetics and Systems Research '92,* R. Trappl (ed). London: World Scientific.

Jibu, M. and Yasue, K. 1993A. Introduction to quantum brain dynamics. In *Nature, Cognition, and System III,* M. E. Carvallo (ed). Amsterdam: Kluwer Academic.

Jibu, M. and Yasue, K. 1993B. Intracellular quantum signal transfer in Umezawa's quantum brain dynamics. *Cybernetics and Systems: An International Journal 24*, 1-7.

Jibu, M. and Yasue, K. 1993C. The basics of quantum brain dynamics. In *Rethinking Neural Networks: Quantum Fields and Biological Data,* K. H. Pribram (ed). New Jersey: Lawrence Erlbaum.

Jibu, M., Hagan, S., Hameroff, S. R., Pribram, K. H. and Yasue, K. 1994. Quantum optical coherence in cytoskeletal microtubules: implications for brain function. *BioSystems 32*, 195-209.

John, E. R. 1972. *Science 177*, 850-864.

John, E. R., Bartlett, F., Shimokochi, M. and Kleinman, D. 1973. *Journal of Neurophysiology 36*, 893-924.

Jung, C. G. and Pauli, W. 1955. *The Interpretation of Nature and the Psyche.* New York: Bollingen Foundation.

Lem, S. 1961. *SOLARIS*. San Diego: Harvest Books.

Nelson, E. 1967. *Dynamical Theories of Brownian Motion*. New Jersey: Princeton University Press.

Nelson, E. 1985. *Quantum Fluctuation*. New Jersey: Princeton University Press.

Pauli, W. 1925. Über den Zusammenhang des Abschlusses der Elektronengruppen im Atom mit der Komplexstruktur der Spektren. *Zeitschrift für Physik 31*, 765-783.

Penrose, R. 1989. *The Emperor's New Mind*. London: Oxford University Press.

Penrose, R. 1994. *Shadows of the Mind*. London: Oxford University Press.

Pribram, K. H. 1991. *Brain and Perception*. New Jersey: Lawrence Erlbaum.

Ricciardi, L. M. and Umezawa, H. 1967. *Kybernetik 4*, 44-48.

Riemann, B. 1854. *Uber die Hypothesen, welche der Geometrie zu Grundeliegen*. (Lecture at Göttingen, June 10).

Sayers, B. McA., Beagley, H. A. and Henshall, W. R. 1974. The mechanism of auditory evoked EEG responses. *Nature 247*, 481-483.

Sayers, B. McA. and Beagley, H. A. 1974. Objective evaluation of auditory evoked EEG responses. *Nature 251*, 608-609.

Schrödinger, E. 1926. Qiantisierung als Eigenwertproblem, I, II and III. *Annalen der Physik 79*, 361-376, 489-527 and *Annalen der Physik 80*, 437-490.

Schrödinger, E. 1944. *What is Life?* Cambridge: Cambridge University Press.

Stuart, C. I. J. M., Takahashi, Y. and Umezawa H. 1978. On the stability and non-local properties of memory. *Journal of Theoretical Biology 71*, 605-618.

Stuart, C. I. J. M., Takahashi, Y. and Umezawa H. 1979. Mixed-system brain dynamics: neural memory as a macroscopic ordered state. *Foundations of Physics 9*, 301-327.

Tonomura, A., Endo, J., Matsuda, T., Kawasaki, T. and Ezawa, H. 1989. Demonstration of single-electron buildup of an interference pattern. *American Journal of Physics 57*, 117-120.

Thomas, L. 1975. *The Lives of a Cell*. New York: Bantam Books.

Tsuchiya, Y., Inuzuka, E., Kurono, T. and Hosoda, M. 1982. Photon-counting imaging and its application. In *Advances in Electronics and Electron Physics 64A*, P. Hawkes (ed). New York: Academic Press.

Umezawa, H. 1993. *Advanced Field Theory: Micro, Macro, and Thermal Physics*. New York: American Institute of Physics.

von Neumann, J. 1966. *Theory of Self-Reproducing Automata*. Urbana: Illinois University Press.

Watterson, J. G. 1987. A role for water in cell structure. *Biochemical Journal 248*, 615-617.

Webb, S. J. 1980. Laser Raman spectroscopy of living cells. *Physics Reports 60*, 201-224.

Webb, S. J. and Stoneham, M. E. 1977. Resonances between 10^{11} and 10^{12} Hz in active bacterial cells as seen by laser Raman spectroscopy. *Physics Letters A60*, 267-268.

Werbos, P. J. 1993. Quantum theory and neural systems: Alternative approaches and a new design. In *Rethinking Neural Networks: Quantum Fields and Biological Data*, K. H. Pribram (ed). New Jersey: Lawrence Erlbaum, and private communication.

Wiener, N. 1948. *Cybernetics: or Control and Communication in the Animal and the Machine*. Cambridge: MIT Press.

Yasue, K. 1978. Quantum mechanics of nonconservative systems. *Annals of Physics (New York) 114*, 479-496.

Yasue, K. 1980. Schrödinger's variational method of quantization revisited. *Letters in Mathematical Physics 4*, 143-146.

Yasue, K. 1981. Quantum mechanics and stochastic control theory. *Journal of Mathematical Physics 22*, 1010-1020.

Yasue, K. 1988. Wave cybernetics: A simple model of wave-controlled nonlinear and nonlocal cooperative phenomena. *Physical Review A38*, 2671-2673.

Yasue, K., Jibu, M., Misawa, T. and Zambrini, J.-C. 1988. Stochastic neurodynamics. *Annals of the Institute of Statistical Mathematics 40*, 41-59.

Yukawa, H. 1935. On the interaction of elementary particles. I. *Proceedings of Physical and Mathematical Society of Japan 17*, 48-57.

Yukawa, H. 1966. Atomistics and the divisibility of space and time. *Progress of Theoretical Physics, Supplement 41*, 512-523.

APPENDICES

Appendix A

Umezawa's Vision: A Modern World View of Quantum Field Theory

This appendix is a brief introduction to a modern world view of physics derived from quantum field theory. The Late Dr. Hiroomi Umezawa, Killam Memorial Professor of Physics at the University of Alberta, Edmonton, Canada, proposed this view in "*Advanced Field Theory: Micro, Macro, and Thermal Physics*." This book, published in 1993 by the American Institute of Physics, is the culmination of his lifelong, original research.

Having resigned his position as the young Professor of Theoretical Physics at the University of Tokyo in 1964, Dr. Hiroomi Umezawa began sowing the seeds of quantum field theory in high energy physics and elementary particle physics in Italy, the United States, and Canada. With a good harvest of quantum field theory in physics, he had energy to spare to develop a modern world view of physics from quantum field theory. Even after the quantum revolution of physics in the 1920s, the conventional world view of physics did not change drastically; it remained the same as the traditional world view of Newton. Namely, reality on our everyday scale is given by macroscopic matter subject to Newton's laws of mechanics, and the atomic constituents of matter in the sub-microscopic scale are subject to fundamental principles of quantum mechanics and quantum field theory. The world view of physics was still divided into two parts: the classical world view for macroscopic phenomena and the quantum world view for sub-microscopic phenomena.

Since the discovery of quantum mechanics, the tacit assumption has been that the quantum world view of physics might reduce to the classical world view by simply passing the Planck constant to the limit of vanishing, that is, $\hbar \to 0$. However, such a naive and simplistic belief cannot be justified, as pointed out by Niels Bohr (Umezawa 1993). Bohr suggested that the quantum world view would reduce to the classical world view only at the limit of an infinite number of atomic ingredients condensed in a spatial region of macroscopic scale. Umezawa turned Bohr's suggestion over and finally succeeded in deriving a modern, unified world view of physics from the fundamental principles of quantum field theory. Umezawa's unified view encompasses both macroscopic objects governed by the phenomenological laws of classical mechanics and sub-microscopic objects governed by the fundamental laws of quantum mechanics.

Thanks to Umezawa's lifelong, enthusiastic interest in clarifying reality in our universe, we can share a drastically new, unified world view of physics in which classical mechanics and field theory for macroscopic objects and quantum mechanics for sub-microscopic objects are mutually derived from quantum field theory. The authors invite you to enjoy this modern world view: Umezawa's vision.

On the 25th of March, 1995, Dr. Hiroomi Umezawa departed from his devoted wife Tamae and ascended to heaven.

Let us take a brief look at the essentials of the modern unified world view of physics in which both macroscopic objects governed by the phenomenological laws of classical physics and sub-microscopic objects governed by the fundamental laws of quantum physics are derived from quantum field theory. Here, macroscopic object means not only macroscopic matter, such as a stone, but also macroscopic fields, such as the electromagnetic field carrying the electromagnetic waves of television, radio, and sunshine.

The basic principles of quantum field theory are described in a way that does not necessarily conflict with Einstein's special theory of relativity; however, we will work with a non-relativistic framework of quantum field theory adequate for the task of investigating low-energy physical phenomena, such as life phenomena. The non-relativistic framework is a specific approximation of quantum field theory, familiar to us in everyday experience, in which the speed of light can be regarded as virtually infinite.

The most fundamental idea of quantum field theory is that reality in our universe is given by certain quantized fields spread over the whole of space. A quantized field $\varphi(\mathbf{x},t)$ in coordinate space has a conjugate representation in momentum space related by a Fourier transformation,

$$\varphi(\mathbf{x},t) = \left(\frac{1}{\sqrt{2\pi}}\right)^3 \int a(\mathbf{k}) e^{-i\omega(\mathbf{k})t} e^{i\mathbf{k}\cdot\mathbf{x}} d^3k \tag{1}$$

where the Fourier transform $a(\mathbf{k})$ of the quantized field $\varphi(\mathbf{x},t)$ is regarded not as an ordinary complex number but as a linear operator acting on vectors in a particular kind of Hilbert space called Fock space and denoted by $\mathfrak{R}$. In the case of a quantized field given not in the whole of space but only in a finite volume, say V, the Fourier transformation is given not by the Fourier integral (1) but by the Fourier series expansion,

$$\varphi(\mathbf{x},t) = \frac{1}{\sqrt{V}} \sum_{\mathbf{k}} a(\mathbf{k}) e^{-i\omega(\mathbf{k})t} e^{i\mathbf{k}\cdot\mathbf{x}} \tag{2}$$

As a linear operator, the Fourier transform $a(\mathbf{k})$ is assumed to satisfy the commutation relations,

$$\left[a(\mathbf{k}),a^*(\mathbf{q})\right] \equiv a(\mathbf{k})a^*(\mathbf{q}) - a^*(\mathbf{q})a(\mathbf{k})$$
$$= \delta^3(\mathbf{k}-\mathbf{q}) \tag{3}$$

where $a^*(\mathbf{q})$ is the adjoint operator of $a(\mathbf{q})$ so that

$$(a^*(\mathbf{q})|\text{vector1}\rangle, |\text{vector2}\rangle) = (|\text{vector1}\rangle, a(\mathbf{q})|\text{vector2}\rangle)$$

holds identically for any vectors $|\text{vector1}\rangle$ and $|\text{vector2}\rangle$ in Fock space $\mathfrak{R}$. Here, (,) denotes the inner product given in Fock space $\mathfrak{R}$ and δ^3 denotes Dirac's delta function. The functional form of $\omega(\mathbf{k})$ may be different for different quantized fields. In other words, $\omega(\mathbf{k})$ characterizes the quantized field $\varphi(\mathbf{x},t)$ together with the Fourier transform $a(\mathbf{k})$. For example, any quantized field given by (1) with

$$\omega(\mathbf{k}) = \frac{\hbar}{2}|\mathbf{k}|^2$$

is called a free quantized field or simply a free field. Other functional forms of $\omega(\mathbf{k})$ give quantized fields called interacting fields.

The quantized fields, whose Fourier transforms are subject to the commutation relations (3), are called Bose-Einstein fields or simply Bose fields. It is possible to impose instead anti-commutation relations, that is,

$$\left[a(\mathbf{k}),a^*(\mathbf{q})\right]_+ \equiv a(\mathbf{k})a^*(\mathbf{q}) + a^*(\mathbf{q})a(\mathbf{k})$$
$$= \delta^3(\mathbf{k}-\mathbf{q}) \tag{4}$$

$$\left[a(\mathbf{k}),a(\mathbf{q})\right]_+ \equiv 0 \tag{5}$$

$$\left[a^*(\mathbf{k}),a^*(\mathbf{q})\right]_+ \equiv 0 \tag{6}$$

If the Fourier transform of any quantized field satisfies these equations, the quantized field is called a Fermi-Dirac field or simply a Fermi field. Although modifications of the commutation relations (3) and the anti-commutation relations (4) are possible, the quantized fields needed to describe all known physical phenomena are thus far either Bose or Fermi fields.

Besides the commutation or anti-commutation relations, the Fourier transform $a(\mathbf{k})$ is assumed to specify a vector $|0\rangle$ in Fock space $\mathfrak{R}$ by the relation

$$a(\mathbf{k})|0\rangle = 0 \tag{7}$$

for all $\mathbf{k}$. This vector $|0\rangle$ is called a vacuum state of the quantized field and is supposed to represent a specific physical state of the quantized field in which no wave motion is present. In other words, it is the minimum energy state of the dynamics of the quantized field. A wave motion of any quantized field is called a quantum of that quantized field. In quantum field theory, such quanta (plural of quantum) are regarded as elementary particles constituting all the physical objects in this universe, e.g., matter and light. The term "vacuum" comes from the fact that there are no physical objects in the vacuum, neither matter nor light.

Suppose we have only one quantized field in this universe, perhaps as a first approximation of our reality, which seems to demand many different quantized fields. Recall that the vacuum state $|0\rangle$ represents a specific physical situation in which no quantum (i.e., no wave motion of the quantized field) exists. Then, starting from the vacuum state $|0\rangle$, a finite series of vectors in Fock space $\mathfrak{R}$ can be specified:

$$\begin{aligned} |\mathbf{k}_1\rangle &\equiv a^*(\mathbf{k}_1)|0\rangle \\ |\mathbf{k}_1\mathbf{k}_2\rangle &\equiv \frac{1}{\sqrt{2!}}a^*(\mathbf{k}_2)a^*(\mathbf{k}_1)|0\rangle \\ |\mathbf{k}_1\mathbf{k}_2\mathbf{k}_3\rangle &\equiv \frac{1}{\sqrt{3!}}a^*(\mathbf{k}_3)a^*(\mathbf{k}_2)a^*(\mathbf{k}_1)|0\rangle \\ &\vdots \end{aligned} \tag{8}$$

To see the meaning of those vectors in Fock space $\mathfrak{R}$, it is convenient to introduce the number operator of the quantized field:

$$N = \int a^*(\mathbf{k})a(\mathbf{k})d^3k \tag{9}$$

Application of the number operator N to the finite series of vectors $|\mathbf{k}_1\rangle$, $|\mathbf{k}_1\mathbf{k}_2\rangle$, $|\mathbf{k}_1\mathbf{k}_2\mathbf{k}_3\rangle$, $\cdots$yields the following relations:

$$\begin{aligned} N|0\rangle &= 0 \\ N|\mathbf{k}_1\rangle &= 1|\mathbf{k}_1\rangle \\ N|\mathbf{k}_1\mathbf{k}_2\rangle &= 2|\mathbf{k}_1\mathbf{k}_2\rangle \\ N|\mathbf{k}_1\mathbf{k}_2\mathbf{k}_3\rangle &= 3|\mathbf{k}_1\mathbf{k}_2\mathbf{k}_3\rangle \\ &\vdots \\ N|\mathbf{k}_1\mathbf{k}_2\mathbf{k}_3\cdots\mathbf{k}_\ell\rangle &= \ell|\mathbf{k}_1\mathbf{k}_2\mathbf{k}_3\cdots\mathbf{k}_\ell\rangle \\ &\vdots \end{aligned} \tag{10}$$

The preceding relations (10) imply that each vector $|\mathbf{k}_1\mathbf{k}_2\mathbf{k}_3\cdots\mathbf{k}_\ell\rangle$ can be interpreted as a dynamical state of the quantized field in which a definite number of quanta (say, ℓ quanta) are present. Therefore, it is customary in quantum field theory to call the vector $|\mathbf{k}_1\mathbf{k}_2\mathbf{k}_3\cdots\mathbf{k}_\ell\rangle$ an ℓ-particle state of the quantized field. From the definition (8) we immediately see that the adjoint operator $a^*(\mathbf{k})$ creates one quantum. Conversely, the Fourier transform $a(\mathbf{k})$ itself annihilates one quantum, because we have

$$N\{a(\mathbf{q})|\mathbf{k}_1\mathbf{k}_2\mathbf{k}_3\cdots\mathbf{k}_\ell\rangle\} = (\ell-1)\{a(\mathbf{q})|\mathbf{k}_1\mathbf{k}_2\mathbf{k}_3\cdots\mathbf{k}_\ell\rangle\} \tag{11}$$

It is in this sense that the Fourier transform of the quantized field, $a(\mathbf{k})$, and its adjoint operator $a^*(\mathbf{k})$ are called annihilation and creation operators, respectively. The relation (7) characterizing the vacuum state asserts that the application of an annihilation operator to the vacuum state, in which no quanta are present, results in zero, that is, a specific vector in Fock space $\mathfrak{R}$ with vanishing length. In quantum field theory, the absolute square of the inner product of any two vectors in Fock space,

$$|(|\text{vector1}\rangle, |\text{vector2}\rangle)|^2$$

represents a probability of realizing a certain physical process. Therefore, the annihilation of a quantum from the vacuum state can be realized with probability zero. This is the meaning of the relation (7).

As far as the physical processes involving a finite number of quanta are concerned, the quantized field $\varphi(\mathbf{x},t)$ remains in any dynamical state which can be represented as a superposition (i.e., weighted sum) of those ℓ-particle states. Such physical processes are sub-microscopic ones involving a finite number of elementary particles. Those processes are realized in any sub-microscopic circumstance; however, they can be observed directly only in the vacuum circumstance of outer space or of an artificially made chamber in a huge accelerator. In those environments, one can look for new elementary particles like quarks. In such circumstances, quantum field theory can be reduced to the more familiar framework of quantum theory, called quantum mechanics (Umezawa 1993). Therefore, the famous conceptual difficulties of quantum mechanics such as "nonlocality," "superposition of states," "the uncertainty principle," "the reduction or collapse of wavefunction (state)," "the EPR-paradox," "non-separability," "Bell's theorem," and "measurement by abstract ego" are all restricted to a sub-microscopic vacuum circumstance, as in outer space or in an experimental apparatus in which the quantum dynamics of a finite number of elementary particles can be directly observed by means of special measurement devices. In such a sub-microscopic vacuum circumstance, all the (finite number of) quanta manifest only sub-microscopic phenomena, which are not directly perceptible to us. We live, however, in a macroscopic world filled with an almost infinite variety of macroscopic objects. It seems natural to doubt the application of quantum mechanics to the investigation of macroscopic phenomena, including, of course, those taking place in the brain. As most brain and cognitive scientists believe, the brain is a typical macroscopic object, and the naive application of quantum mechanical concepts to brain functioning cannot be justified. However, quantum field theory is so fundamental that it encompasses much more than quantum mechanics.

The use of quantum mechanics in investigating macroscopic objects is untenable. Nevertheless, we should not go back to good old classical physics. We need a deeper understanding based on the first principles of theoretical physics. Classical physics provides us with only a phenomenological understanding. We have to look for a new theoretical framework. Fortunately, Umezawa showed in 1993 that the basic principles of quantum field theory are sufficiently rich and fundamental to derive not only the quantum mechanics of sub-microscopic phenomena but also a fundamental framework for the investigation of macroscopic phenomena. This framework constitutes the major part of quantum field theory; quantum mechanics is only a minor part of it. So it seems natural to call this major part also "quantum field theory."

Let us see in what way quantum field theory opens the possibility of describing and investigating macroscopic phenomena, once studied only phenomenologically within the realm of classical physics (i.e., classical mechanics and classical field theory). The key concept is given by the Bogoliubov transformation of the quantized field: First, the Bogoliubov transformation of the creation and annihilation operators is defined as

$$a(\mathbf{k}) \to \alpha(\mathbf{k}) \equiv a(\mathbf{k}) + \theta(\mathbf{k}) \tag{12}$$

$$a^*(\mathbf{k}) \to \alpha^*(\mathbf{k}) \equiv a^*(\mathbf{k}) + \overline{\theta}(\mathbf{k}) \tag{13}$$

where $\theta(\mathbf{k})$ is a given complex-number-valued function of $\mathbf{k}$ and $\overline{\theta}(\mathbf{k})$ denotes its complex conjugate. Then it is immediately seen that the linear operators $\alpha(\mathbf{k})$ and $\alpha^*(\mathbf{k})$ acting on vectors in Fock space $\mathfrak{R}$ also satisfy the commutation relations:

$$\begin{aligned} \left[\alpha(\mathbf{k}), \alpha^*(\mathbf{q})\right] &\equiv \alpha(\mathbf{k})\alpha^*(\mathbf{q}) - \alpha^*(\mathbf{q})\alpha(\mathbf{k}) \\ &= \delta^3(\mathbf{k}-\mathbf{q}) \\ \left[\alpha(\mathbf{k}), \alpha(\mathbf{q})\right] &= 0 \\ \left[\alpha^*(\mathbf{k}), \alpha^*(\mathbf{q})\right] &= 0 \end{aligned} \tag{14}$$

Of course, this is the case of a Bose field, and that of a Fermi field requires a slight modification. Namely, $\theta(\mathbf{k})$ and $\overline{\theta}(\mathbf{k})$ are no longer complex-number-valued functions of $\mathbf{k}$, but Grassmann-number-valued functions subject to the anti-commutation relations:

$$[\theta(\mathbf{k}), \theta(\mathbf{q})]_+ = 0$$

$$[\overline{\theta}(\mathbf{k}), \overline{\theta}(\mathbf{q})]_+ = 0$$

Then, the linear operators given by the Bogoliubov transformations (12) and (13) come to satisfy the anti-commutation relations (4)-(6) with a replaced by α.

Secondly, since the transformed linear operators α and α^* are subject to the same commutation or anti-commutation relations as the creation and annihilation operators, they can be used to characterize another vacuum state $|0(\theta)\rangle$ of the quantized field $\varphi(\mathbf{x},t)$ through this relation:

$$\alpha(\mathbf{k})|0(\theta)\rangle = 0 \tag{15}$$

As we can choose any functional form for $\theta(\mathbf{k})$ from among the continuously infinite possibilities, there exists a continuously infinite variety of vacuum states of the quantized field. The vacuum state $|0\rangle$, which played an important role in describing the sub-microscopic dynamics of a finite number of quanta in the sub-microscopic scale, is now found to be only one possibility, characterized by vanishing θ among infinitely many candidates for the vacuum state. Then, a few questions naturally may arise: What are these infinitely many vacuum states? What kind of physical situations do they correspond to? Are they all equivalent to each other? Those questions indeed triggered the modern unified world view of physics within the realm of quantum field theory.

It will be convenient to make explicit the form of the vacuum state $|0(\theta)\rangle$. Call it a θ-vacuum of the quantized field φ:

$$|0(\theta)\rangle = e^{-\frac{1}{2}\int |\theta(\mathbf{k})|^2 d^3k} e^{\int \theta(\mathbf{k}) a^*(\mathbf{k}) d^3k} |0\rangle \tag{16}$$

This equation may be interpreted as defining the Bogoliubov transformation of the specific vacuum state $|0\rangle$, call it the true vacuum, into the θ-vacuum. It is astonishing that the θ-vacuum is orthogonal to all of the finite-number-particle states in the sense that the inner products of the θ-vacuum with each of the finite-number-particle states (including the zero-particle state, that is, the true vacuum state) all vanish:

$$(|\mathbf{k}_1\mathbf{k}_2\mathbf{k}_3\cdots\mathbf{k}_\ell\rangle, |0(\theta)\rangle) = 0$$

This fact can be interpreted as follows: since the inner product of the θ-vacuum with every finite-number-particle state vanishes, there is zero probability of the possibility that the θ-vacuum represents a physical situation of the quantized field φ manifesting any finite number of quanta (including zero quanta, of course). It is a dynamical state of the quantized field, which cannot be any finite-number-particle state; it cannot even be the true vacuum state. Therefore, the quantized field φ must be regarded as containing infinitely many quanta in the θ-vacuum. Such a dynamical state of the quantized field with infinitely many quanta in the θ-vacuum is called a collective mode, or macroscopic quantum state, or condensed state, or simply a condensate.

The Bogoliubov transformation of the creation and annihilation operators defined by (12) and (13) induces a transformation of the quantized field $\varphi(\mathbf{x},t)$ through the Fourier transformation (1) or (2), with this result:

$$\varphi(\mathbf{x},t) \to \tilde{\varphi}(\mathbf{x},t) \equiv \left(\frac{1}{\sqrt{2\pi}}\right)^3 \int \alpha(\mathbf{k}) e^{-i\omega(\mathbf{k})t} e^{i\mathbf{k}\cdot\mathbf{x}} d^3k \tag{17}$$

This is called the Bogoliubov transformation of the quantized field $\varphi(\mathbf{x},t)$. Equations (1) and (12) yield

$$\begin{aligned}\tilde{\varphi}(\mathbf{x},t) &\equiv \left(\frac{1}{\sqrt{2\pi}}\right)^3 \int a(\mathbf{k}) e^{-i\omega(\mathbf{k})t} e^{i\mathbf{k}\cdot\mathbf{x}} d^3k + \left(\frac{1}{\sqrt{2\pi}}\right)^3 \int \theta(\mathbf{k}) e^{-i\omega(\mathbf{k})t} e^{i\mathbf{k}\cdot\mathbf{x}} d^3k \\ &= \varphi(\mathbf{x},t) + \lambda(\mathbf{x},t)\end{aligned} \tag{18}$$

where $\lambda(\mathbf{x},t)$ is nothing but a complex-number-valued (or real-number-valued) function of position and time whose Fourier transform is given by $\theta(\mathbf{k})$. In the case of a Fermi field, $\lambda(\mathbf{x},t)$ becomes a Grassmann-number-valued function of position and time, of course. In quantum field theory, such a complex-number-valued or Grassmann-number-valued function of position and time is regarded as a macroscopic field, like Maxwell's electromagnetic field, a classical field investigated by classical field theory. Thus, the Bogoliubov transformation of the quantized field $\varphi(\mathbf{x},t)$ results simply in a superposition of the quantized field itself with a classical field $\lambda(\mathbf{x},t)$.

Combining this fact with the aforementioned property of the θ-vacuum obtained also by the Bogoliubov transformation, the transformed quantized field $\tilde{\varphi}$ can now be interpreted as representing a macroscopic object, such as macroscopic matter and light, made of infinitely many quanta: Any macroscopic object is given by a superposition of a (macroscopic) classical field, representing the mean (macroscopic) envelope structure of the macroscopic object, and a (sub-microscopic) quantized field, representing the fluctuating (sub-microscopic) inner structure of the macroscopic object. The former contains infinitely many quanta; thus, it is independent of the detailed characteristics of the dynamics of those quanta, thanks to the law of large number. The situation is similar to our actual experience in the macroscopic world. Indeed, a small stone can be thrown by hand, regardless of the detailed characteristics of the dynamics of the infinitely many electrons, protons, and neutrons composing the stone!

The fluctuating (sub-microscopic) inner structure of the macroscopic object is represented by the original quantized field φ, so it is essentially equivalent to the sub-microscopic dynamics of a finite number of quanta well described by quantum mechanics. Once again, the situation closely matches our actual one in the macroscopic world. The fluctuating (sub-microscopic) inner structure of the (macroscopic) stone must be investigated not by the phenomenology of classical mechanics but by the fundamental laws of quantum mechanics.

With the help of the Bogoliubov transformation of the quantized field, we have shown succinctly that any macroscopic object, such as macroscopic matter and light, can be well described by the θ-vacuum manifesting infinitely many quanta. A good old phenomenological description of a macroscopic object as a classical material body or a classical field in classical physics is derived as the mean (macroscopic) envelope structure of the θ-vacuum containing infinitely many quanta. Simultaneously, the well-known fundamental framework of quantum mechanics for the atomic constituents of the macroscopic object is derived as the fluctuating (sub-microscopic) inner structure of the θ-vacuum, essentially the same as the quantum mechanics of a finite number of quanta in the sub-microscopic vacuum circumstance of outer space or of the accelerator chamber, but a slight modification is required because the system of a finite number of quanta in the condensed circumstance of infinitely many quanta constituting the macroscopic object suffers from complicated perturbations caused by the mean (macroscopic) envelope structure of the θ-vacuum containing infinitely many quanta.

Thanks to the law of large number, one can treat such perturbations as a random variable. Furthermore, one can find at least two approximations to describe the quantum mechanics of a system of a finite number of quanta in the condensed circumstance: Quantum statistical mechanics is a phenomenological approximation based on Gibbs' hypothesis of the equivalence of the long-time average to the fictitious "ensemble" average. Thermo-field dynamics (Umezawa 1993) is a truly quantum field theoretical approximation based on random creation and annihilation operators and the Bogoliubov transformation.

The latter succeeded in describing the dynamics of quanta in the condensed circumstance without recourse to any phenomenology, so it seems plausible for future development. Both classical physics for macroscopic objects and quantum physics for a finite number of quanta (not only in the vacuum circumstance but also in the condensed circumstance of infinitely many quanta constituting the macroscopic object) are derived theoretically from the basic principles of quantum field theory.

The modern unified theory of physics derived from the basic principles of quantum field theory is capable of describing and investigating both sub-microscopic and macroscopic phenomena simultaneously. To proceed into the terra incognita of "consciousness," one must adopt, not the old phenomenology of classical physics, but a new and more fundamental point of view: one based on a modern unified theory of physics derived from quantum field theory by Hiroomi Umezawa.

Appendix B

Glossary

Cross-references are indicated in bold.

A

absolute square

A complex number can be always expressed as a sum: $z=x+iy$, where i is the **imaginary unit number** so that $i^2=-1$, and x and y are two real numbers, called the real part and imaginary part of a complex number, respectively. An absolute square of a complex number $z=x+iy$ is a real number indicating the magnitude or norm of a complex number z given by the sum x^2+y^2.

absolute value

An absolute value of a complex number $z=x+iy$ is a real number indicating the magnitude or norm of a complex number z given by the square root of the sum x^2+y^2.

acceleration

The change in **velocity** in both magnitude and direction of **matter** with respect to time is called the acceleration.

amplitude

In the transfer of **energy** by a **wave** form of displacement of either a material medium or a **field**, the maximum displacement is called amplitude.

atom

The atom is the smallest unit of **matter** stable against any chemical reaction and composed of several kinds of **elementary particles**. Its central core, the **nucleus**, consists of positively charged particles, called **protons**, and uncharged (electrically neutral) particles, called **neutrons**. Surrounding the nucleus are negatively charged particles, called **electrons**. Each atom has an equal number of protons and electrons, so it remains uncharged as a whole.

automaton

An automaton is an apparatus that automatically performs certain actions by responding to present conditions or encoded instructions; it acts in a mechanical way. An automaton requires a programmed set of instructions to perform a function.

Avogadro number (Avogadro's number)

A number 6.02252 times 10^{23} is called the Avogadro number or Avogadro's number. Macroscopic matter contains approximately the Avogadro number of **atoms** or **molecules.**

B

Bohm, David

Physicist David Bohm was born in the United States of America in 1917 and passed away in England in 1992. The most valuable contribution he made to physics was the restoration of the **ontological interpretation of quantum mechanics** due originally to **de Broglie.** Bohm did not confine himself to his own restricted domain of physics. He also proposed a completely new world view modeled upon his **ontological interpretation of quantum mechanics.** This view serves as the theoretical basis of the holographic paradigm put forth by Karl Pribram. Bohm's work has significantly influenced late twentieth century philosophy.

Bohr, Niels

Danish physicist Niels Bohr solved a problem that **classical physics** had been unable to explain: the stability of the **atom** made of **electrons** surrounding the **nucleus. Newton's classical mechanics** and **Maxwell's theory of electromagnetism** led to the conclusion that the atom would collapse by losing **energy**, due to the continuous emission of **electromagnetic waves** from **electrons** moving around the **nucleus.** Bohr overcame this difficulty by postulating that **electrons** move in restricted orbits around the **nucleus** and do not emit **electromagnetic waves** while remaining in each restricted orbit. The condition satisfied by those restricted orbits is called Bohr's quantum condition.

Boltzmann distribution (Maxwell-Boltzmann distribution)

The **probability** distribution of **velocity** and the position of **molecules** of ideal gas in thermal equilibrium was obtained originally by **James Maxwell** and generalized later by Ludwig Boltzmann within the framework of **classical mechanics.**

Bose-Einstein condensate (Bose condensate)

The minority of **quanta** constituting **matter** in the natural world manifest a specific characteristic of **quantum theory** discovered by Indian physicist Satyendra Nath Bose and then promoted by **Albert Einstein.** Namely, each **quantum state** (of motion) for those particular **quanta** can be occupied by any number of **quanta**, and extremely many **quanta** can be in the same **quantum state.** The **quantum state** (of motion) occupied by extremely many **quanta** is called a Bose-Einstein condensate.

Bose-Einstein condensation (Bose condensation)

The phenomenon in which a **Bose-Einstein condensate** emerges is called Bose-Einstein condensation.

C

classical mechanics (Newtonian mechanics)

The branch of physics concerned with the motion of **matter** is called mechanics. **Isaac Newton** discovered three laws of motion and the law of universal gravitation. He set forth a theoretical framework of mechanics based on those laws in his book *PHILOSOPHIE NATURALIS PRINCIPIA MATHEMATICA* in 1687. Mechanics with this theoretical framework was called Newtonian mechanics or, more simply, mechanics, but designated as "**classical physics**" in the 1920s when **quantum mechanics** superseded Newton's framework.

classical physics

Classical physics refers to the traditional branches of physics developed before the dawn of **quantum theory** at the beginning of the twentieth century. Classical physics, based primarily on **Newtonian mechanics** and **Maxwellian electromagnetism**, pertains to natural phenomena of **matter** and **fields** on the normal (macroscopic) scale of observation.

collective unconscious

A term used by Swiss psychiatrist **Carl Jung** to indicate a **consciousness** that contains symbols and archetypes common to all human beings.

complex plane (Gauss plane)

A fictitious plane, that is, a two dimensional space with rectangular coordinates x and y, is called a complex plane if each point on the plane with coordinates (x,y) is identified with a complex number $z=x+iy$, where i is the **imaginary unit number**. As the complex plane was used first by the German mathematician, Carl Friedrich Gauß, it is also called a Gauss plane.

consciousness

Consciousness is a term indicating the totality of mental phenomena a human being experiences.

conservation law

The basic laws in physics maintaining that the total value of certain quantities remains unchanged during a physical process are called conservation laws. Among conserved quantities are **mass**, **energy**, **momentum**, and electric charge.

conservation of probability (continuity of probability)

The conservation of probability, also termed the continuity of probability, pertains to the fact that the increment of **probability** of finding something in any small region of space is always equal to the difference between the probability flowing into the region and the probability flowing out of the region. The difference in question is given by the **divergence** of the **probability current**. The conservation of probability also means that the probability of finding something in the whole of space is equal to one.

cosmic ray

A ray of high-energy **elementary particles**, such as **protons**, bombarding the earth from outer space is called a cosmic ray.

D

de Broglie, Louis

French physicist Louis de Broglie hypothesized in 1923 that the fundamental constituent of **matter** exhibits not only a particle-like property but also a wave-like property, and he assumed the existence of **waves** behind **matter**, called **matter waves**. His essential contribution to the development of **quantum mechanics** led to his award of the 1929 Nobel Prize in physics. He also originated the **ontological interpretation of quantum mechanics** in terms of the pilot wave.

degree of freedom

The fundamental ability or capacity of **matter** or a **field** to change its configuration is called the degree of freedom.

Descartes, René

French philosopher, mathematician, and physicist René Descartes is considered the father of modern philosophy and science. He originated a mechanistic world view in which the **law of causality** played an essential role.

Dirac, Paul Adrien Maurice

English physicist Paul Dirac discovered the fundamental **wave** equation of **quantum mechanics**, later called the Dirac equation, which superseded the **Schrödinger equation** when the **special theory of relativity** was taken into account. He contributed also to the development of **quantum field theory** and quantum electrodynamics and elucidated the nature of the Fermi-Dirac **field** and **quantum**. He is recognized as one of the greatest physicists of all time, along with **Einstein** and **Schrödinger**.

divergence

The total difference between all the outward pointing **vectors** and all the inward pointing vectors in a vector field on a point in space is called the divergence of the **field**. The term applies to any **field** of **vectors** that extends across a spatial region or the whole of space.

E

ego

The part of the human mind that has a sense of individuality or self-awareness. The "subject" in human consciousness.

Einstein, Albert

Physicist Albert Einstein was born in Germany in 1879 and passed away in the United States of America in 1955. He discovered three revolutionary theories of physics in 1905 when he was an unsung physicist working in the Swiss Patent Office in Bern. They were the statistical theory of Brownian motion from which the existence of the fundamental elements of **matter**, called **molecules**, was first verified, the **quantum theory of light**, and the **special theory of relativity**.

Einstein's Equation

The fundamental equation in the **general theory of relativity** discovered by **Albert Einstein** that determines the large-scale structure of space and time with respect to the distribution of **energy** and **matter** in our universe. Conversely, the distribution of matter and energy is determined by Einstein's equation with respect to the large scale structure of space and time.

electric dipole (electric dipole moment)

A pair of positive and negative electric charges of equal amount situated a small distance apart is called an electric dipole. The strength of an electric dipole given by the distance between the positive and negative charges times the value of the positive charge is called electric dipole moment. An electric-dipole-moment **vector**, that is, an arrow, points from the negative charge to the positive charge; the arrow length is given by the electric dipole moment.

electromagnetic field

A physical object extended across the whole of space causing all the phenomena associated with the **theory of electromagnetism** is called the electromagnetic field. It is a composite **field** of the electric field and the magnetic field, both of which may be seen as **vectors** or rays distributed across the whole of space.

electromagnetic wave

The transfer of **energy** by the variation of intensity of the distributed **vectors** of the **electromagnetic field** in the form of a regular vibration is called an electromagnetic wave. ("Regular" in this context means that the vibration exhibits regularity.)

electron

The most fundamental **elementary particle** carrying a unit charge of negative electricity is called an electron. In **quantum theory**, electrons are thought of as **quanta** of a **field** described by the **Schrödinger** equation or the **Dirac** equation.

elementary particle

The tiniest bits of **matter** or light assumed to be the most basic constituents of this universe are called elementary particles. They are all found to be **quanta** of certain **matter fields** or of the **electromagnetic field** within the realm of **quantum field theory**. Elementary particles are simply said to be **particles**, though they cannot be well understood by a naive picture of the **particle** in **classical physics**, that is, as a small material body.

energy

The ability or capacity of **matter** or a **field** to affect other matter or fields. That ability derived from its position or configuration is called potential energy, and that derived from its motion or change is called kinetic energy. Energy is subject to the conservation law; that is, it can be neither created nor annihilated, but can only change from one form of energy into another form.

entropy

A physical variable or parameter indicating the magnitude of disorder of the dynamics.

ether (aether)

This hypothetical medium for transmitting an **electromagnetic wave** was believed to exist until 1905 when **Einstein** discovered the **special theory of relativity**, which eliminated the need for such a medium. Thereafter, the term was used only in a historical context until 1951 when English physicist **Paul Dirac** claimed that the **special theory of relativity** could not eliminate ether if **quantum theory** were seriously taken into account.

Euler equation

The equation of motion for the perfect fluid, that is, an ideal fluid without viscosity, was found by a Swiss mathematician Leonhard Euler and called the Euler equation. It is nothing but **Newton's law of motion** written in terms of the **field** of the **velocity vector** for a small element of fluid.

exclusion principle

The majority of **quanta** constituting **matter** in the natural world are subject to a specific principle of **quantum theory** discovered by Austro-American physicist **Wolfgang Pauli.** The exclusion principle, as it is called, claims that each **quantum state** (of motion) for those quanta, called Fermi-**Dirac** quanta, can be occupied by only one quantum, and no other can be in the same quantum state.

F

field

A set of infinitely many **degrees of freedom** extended across a spatial region or the whole of space is called a field. The field was first used to describe all the natural phenomena involving **matter** distributed and separated in space but affecting each other as a result of the **law of causality**. **Classical mechanics** and **quantum mechanics** specialized to investigate the field are called classical field theory and **quantum field theory**, respectively.

frequency

The frequency of a **wave** is the number of crests or troughs (representing the maximum displacement of the wave) that pass a point in space per unit of time. The inverse of the frequency of a wave is called the period of a wave.

G

Gauss plane

See **complex plane.**

Gaussian probability law

The probability law governing a **Gaussian random variable** is called the Gaussian probability law.

Gaussian random variable

The sum of many **random variables** of any kind becomes a specific random variable characterized by the mean value and the standard deviation. This specific random variable is called a Gaussian random variable since it was first investigated systematically by the German mathematician Carl Friedrich Gauß in relation to the theory of errors.

general theory of relativity (theory of general relativity)

Based on the principle of equivalence of gravity and acceleration and the **principle of the constancy of the velocity of light**, **Albert Einstein** generalized his **special theory of relativity** and discovered a unified theory of **matter**, gravity, space, and time in **classical physics** in 1916. It is called the general theory of relativity or the theory of general relativity. Here, **Einstein's equation** determines the large-scale structure of space and time with respect to the distribution of **energy** and matter in our universe. Conversely, the distribution of matter and energy is determined by Einstein's equation with respect to the large-scale structure of space and time.

gradient

See **space derivative.**

I

imaginary unit number

A fictitious number whose square equals minus one. It is customary in science to denote the imaginary unit number by the lower case i, so we have $i^2 = -1$.

implicate order

The order which is not apparent in the dynamics of a physical system is called the implicate order. See **David Bohm.**

interference

The effect obtained when two systems of **waves** reinforce or neutralize each other is called interference. When two waves are in the same **phase**, constructive interference occurs and results in a combined wave with the **amplitude** equal to the sum of the amplitudes of the two waves. When two waves are in opposite phases, that is, when the maximum positive amplitude of one wave coincides with the maximum negative amplitude of the other wave, destructive interference occurs and results in the canceling of the waves if they have the same amplitude. In 1804, **Thomas Young** observed the interference effect in his experiments with light. Young's findings prompted the idea that light is a wave; first light was perceived as a wave of **ether** and afterward as a wave of the **electromagnetic field**.

J

Josephson, Brian

English physicist Brian Josephson discovered the anomalous effect of **superconducting current**, later called the **Josephson effect**, for which he received the Nobel prize in physics in 1973 when he was thirty-three years old.

Josephson effect

The sandwich structure of two **superconducting media** separated by a thin layer of insulator is called the Josephson junction. **Brian Josephson** investigated the nature of **superconducting currents** across this junction from the point of view of **quantum mechanics of many-body systems**. Contrary to ordinary matter, the superconducting current across the Josephson junction vanishes if a constant voltage is applied to it, but it does not vanish if an alternative AC voltage is applied.

Jung, Carl Gustav

Swiss psychiatrist Carl Jung founded the analytical school of psychology with a broadened focus on the unconscious. He assumed the unconscious to be composed of a personal unconscious and the collective unconscious. The former is the totality of an individual's repressed or forgotten mental and material experience; the latter is a **consciousness** containing symbols and archetypes common to all human beings. The incorporation of the collective unconscious into psychiatric study led him to a deeper understanding of the phenomenon of acausal occurrence of events that indicate a meaning in their convergence, that is, meaningful coincidence, which Jung termed **synchronicity**.

L

law of causality

The most fundamental law of physics claiming that every phenomenon in the natural world is caused by other phenomena; it is the law of cause and effect.

light quantum (photon)

Albert Einstein proposed the **quantum theory of light** in 1905 to explain the photoelectric effect in which light itself consists of cryptic states of the **wave** motion of the **electromagnetic field** with restricted values of **energy** that behave like **particles** in energy exchange. Those states of the wave motion were called light quanta or photons.

local physical law

A law or principle in physics expressed by either an equation or a relationship among physical variables in neighboring points of space is called a local physical law. **Fields** are known to be subject to local physical laws.

M

mass

The most basic quantity of **matter**, which is nothing but a measure of the resistance of matter to acceleration by an external force.

matter

Anything in this universe is called matter if it is made of **quanta** of **matter fields.**

matter field

The **elementary particles** constituting **matter** in this universe are all **quanta**, that is, specific amounts of **energy** of certain **fields** which can be described only by **quantum theory**. Those fields are called matter fields.

matter wave

Waves of the **matter field** are called matter waves; they manifest the wave-like property of **quanta** of the matter field.

Maxwell, James

The Scottish physicist James Maxwell developed the theory of the **electromagnetic field** on a mathematical basis and concluded that **energy** travels in

electromagnetic waves propagating at a speed equal to that of light, so light is only a type of electromagnetic wave.

Maxwell-Boltzmann distribution

See **Boltzmann distribution.**

memory

Memory is one of the mental phenomena of **consciousness**; it records previous actions, events, and states of being.

mind

Mind, a term indicating the subject who experiences directly the mental phenomena of human **consciousness**, is the action of consciousness.

mind-body dualism

A system of thought explaining all phenomena in terms of two distinct principles—**mind** and **matter**—is called mind-body dualism.

moire

A *moire* pattern is a ripple. The adjective "moire" is derived from French.

molecular biology

A branch of modern biology in which all the biological processes in living matter are investigated from the point of view of chemical reactions or the interaction of biomolecules.

molecule

The smallest **particle** of **matter** that has all the chemical properties of the matter in question is called a molecule. It is made up of one, two, or more **atoms** either of the same kind or of different kinds. The molecules of living matter are called biological molecules or biomolecules.

momentum

The quantity of motion of **matter** given by the product of its **mass** and **velocity** is called momentum. Momentum is conserved if matter is subject to no external action.

N

negentropy

Negentropy is the opposite of **entropy**, a physical variable or parameter indicating the magnitude of disorder of dynamics. Negentropy is a physical variable indicating the magnitude of order of dynamics.

neutron

The uncharged **elementary particle** of slightly greater **mass** than the **proton** is called the neutron. In the past, the neutron and the proton were thought to be two different states of the same elementary particle, the nucleon, but today the proton and the neutron are assumed to be made of three other elementary particles called quarks.

Newton, Sir Isaac

The English mathematician and physicist Sir Isaac Newton discovered three laws of motion, now called **Newton's law of motion**, and the law of universal gravitation. He also devised a theoretical framework of mechanics based on those laws as well as the mathematical framework of differential and integral calculus; he developed those frameworks in his book *PHILOSOPHIE NATURALIS PRINCIPIA MATHEMATICA* in 1687. This theoretical framework of mechanics was called Newtonian mechanics or, more simply, mechanics, but it came to be known as **classical mechanics** after being superseded by **quantum mechanics** in the 1920s.

Newton's law of motion

The set of three laws of motion of **matter** in the natural world discovered by **Newton** is called Newton's law of motion. The first law of motion claims that matter at rest or in uniform, linear motion keeps its state of motion as long as no external action is exerted on it. That is also called the law of inertia. The second law of motion claims that the change of motion of matter measured by acceleration is proportional to the external action measured by force. The second law is frequently called Newton's equation of motion. The third law of motion claims that matter subject to the external action of other matter gives the reaction back to that matter; the action and reaction are the same in strength but opposite in direction.

nonlocality

The **law of causality** asserts that the cause of any effect in the natural world can be localized, that is, specified in an arbitrarily small region of space. That law, which is the most fundamental fact in **classical physics**, is called locality; however, it is no longer valid in **quantum physics**. There exist certain phenomena in which the cause of a quantum theoretical effect cannot be localized but is distributed over a region. This turns out to be the most fundamental fact in quantum physics, and it is called nonlocality.

nuclear force

The binding force between **protons** and **neutrons** in the **nucleus** is called nuclear force.

nucleus

The central core of an **atom** is called the nucleus. It consists of positively charged **particles**, called **protons**, and uncharged (electrically neutral) particles, called **neutrons**. The nucleus occupies only a tiny fraction of the volume of an atom but contains almost all of its **mass**. It is the nucleus that specifies the nature of **matter**. All the atoms of an element of matter have the same number of protons in their nuclei. The protons and neutrons that form a nucleus were once supposed to be two different states of the same **elementary particles**, called nucleons.

O

ontological interpretation of quantum mechanics

In **quantum mechanics** the most fundamental role is played by the wavefunction which, ironically, is a conventional mathematical concept that has no realistic meaning in the mainstream interpretation of quantum mechanics, that is, the Copenhagen interpretation. Although most of the physicists accepted the conventional interpretation of quantum mechanics, a few physicists—including **Albert Einstein**, **Louis de Broglie**, and **David Bohm**—offered a much more realistic interpretation of **quantum mechanics**. Theirs is called the ontological interpretation of quantum mechanics. There, the wavefunction has a realistic meaning: it represents a certain **field** that actually exists and spans over the whole of space. The field is called a ghost field by Einstein, a pilot wave by de Broglie, and a quantum potential by Bohm.

P

particle

Tiny bits of **matter** are called particles. In **classical physics**, they are supposed to be small material bodies subject to **Newton's law of motion**. In **quantum physics**, they are found to be **elementary particles**, which are **quanta** of their corresponding **fields**.

Pauli, Wolfgang

Austro-American physicist Wolfgang Pauli contributed to the theoretical development of both **quantum mechanics** and **quantum field theory** in three significant ways: he discovered the **exclusion principle**; he presented the theoretical formulations of quantum field theory and quantum electrodynamics; and he posited the existence of a new **elementary particle**, called the neutrino today.

phase

The spatiotemporal change of the **amplitude** of a **wave** in space can be decomposed into a periodic motion described by a trigonometric function, that is, a sine function of some function of position and time, multiplied by another function of position and time. The former function is called a phase of the wave, and the latter function is called an envelope of the wave.

photon (light quantum)

See **light quantum**.

Planck constant

One of the most fundamental constants in physics, assumed first by **Max Planck**, was the constant of proportionality between the **energy** emitted or absorbed by an **atom** and the **frequency** of emitted or absorbed **light** as an **electromagnetic wave**.

Planck, Max

The German physicist Max Planck put forth a drastic hypothesis, called the quantum hypothesis, that **atoms** emit and absorb **energy** in terms of **light** only in discrete bundles, called **quanta**, instead of continuously as assumed in **classical physics**.

positron

The anti-particle of an **electron** is called a positron. It is an **elementary particle** with the same **mass** as an electron but with an opposite sign of electric charge, that is, a positive charge. A positron can be interpreted as an electron going from the future to the past.

principle of the constancy of the velocity of light

One of the fundamental hypotheses of **Einstein's special theory of relativity** is that the measured speed of light is constant for all observers, regardless of the motion of the observer or of the source of light. Called the principle of the constancy of the velocity of light, it was later verified by the Michelson-Morley experiment. Acceptance of the principle of the constancy of the velocity of light meant renouncement of the concept of **ether**.

probability

A parameter or variable that measures the likelihood that an event will occur is called a probability. The mathematical theory of probability was originally proposed by Blaise Pascal in the seventeenth century and further developed by Pierre de Laplace in the nineteenth century. It was completed by the Russian mathematician Andrei Kolmogorov in 1933.

probability current

The rate of time change of probability distribution (i.e., the spatial distribution of **probability**) is given by the **divergence** of the probability current represented by a **velocity vector** of a flow of elements of probability distributed in space.

proton

The **nucleus**, that is, the central core of the smallest and lightest **atom**, hydrogen, is assumed to be an **elementary particle** called a proton. It is almost as heavy as a hydrogen atom and has a positive electric charge, which cancels out with the negative electric charge of an **electron**. The **neutron** and proton were once thought to be two different states of the same elementary particle, the nucleon, but today they are assumed to be made of three other elementary particles called quarks.

Q

quanta

The plural form of **quantum**.

quantum

The basic form of the most fundamental constituents of **matter** is a cryptic state of the **wave** motion of the **matter field** with a restricted value of **energy**. It is called a quantum and can be understood as a **particle** from the point of view of energy exchange and as a wave from the point of view of dynamics. In this sense, a quantum is frequently said to bear the complementary characters of wave and particle, the wave-particle dualism. The motion of a quantum can be investigated through **quantum theory**. **Quantum mechanics** deals with low energy processes, and **quantum field theory** deals with high energy processes. All **elementary particles** constituting matter in this universe are quanta.

quantum coherence

In the **quantum mechanics of many-body systems**, some **quanta** can be strongly correlated; they manifest the same motion simultaneously even though they are not interacting directly with each other. Those quanta are said to have quantum coherence. For example, quanta taking part in the **superconducting**

current have quantum coherence.

quantum field theory

Quantum theory specialized to describe the mechanics of the **field** is called quantum field theory.

quantum mechanics

Quantum theory specialized to describe the mechanics of a small number of **elementary particles** is called quantum mechanics.

quantum mechanics of many-body systems

Quantum theory specialized to describe the mechanics of an indefinite number of **elementary particles** is called the quantum mechanics of many-body systems.

quantum physics

All the branches of physics born from **quantum theory** fall within quantum physics.

quantum state

Unlike the notion of the **particle** in **classical physics**, the **quantum** cannot manifest arbitrary motion but only a restricted class of realizable motions. A quantum manifesting restricted motion is said to be in a quantum state (of motion).

quantum theory

A revolutionary theory of physics discovered in the decade of the 1920s, quantum theory claimed that the fundamental constituent of **matter** is not a **particle**, as had been widely believed, but a **quantum** that manifests as a particle or as a **wave** depending on the way it is observed.

quantum theory of light

Albert Einstein proposed a revolutionary theory in 1905 to explain the photoelectric effect. He proposed that light consists of cryptic states of the **wave** motion of the **electromagnetic field** with restricted values of **energy** that behave like **particles** in energy exchange. Those states of the wave motion were called **light quanta** or **photons**, and Einstein's theory was called the quantum theory of light.

quantum tunneling

In **classical mechanics** a **particle** with **energy** less than the value needed to climb over the wall can neither surmount it nor go through it. In **quantum**

theory, however, a **quantum** with energy less than the required value to climb over the wall can go through it because the **probability** of finding the quantum outside the wall given by the **absolute square** of a wavefunction remains non-zero. This is called a quantum tunneling phenomenon.

R

random variable

A physical variable suffering from the probability law is called a random variable. It is a variable whose dynamics cannot be well described without recourse to the concept of **probability**.

S

Schrödinger, Erwin

Austrian physicist Erwin Schrödinger developed in 1926 the most powerful form of **quantum mechanics**, called wave mechanics, in terms of the wavefunction and a wave equation, called the Schrödinger equation today. He shared with **Paul Dirac** the 1933 Nobel Prize in physics.

soliton

A solitary wave, described by a nonlinear wave equation, that maintains its **wave** form while in motion is called a soliton. A soliton is realized as the result of nonlinear and global (long-range) interaction in the background **field** or medium, so it is an expression of global properties.

space derivative (gradient)

A **vector** composed of three components given by the derivatives or differentials with respect to the three coordinates of space x, y, and z is called a space derivative or a gradient.

special theory of relativity

Based on the **principle of the constancy of the velocity of light** and the hypothesis that the laws of the natural world are the same in different moving frames or systems of reference, **Albert Einstein** introduced a revolutionary theory of **classical physics** in 1905. It is called the special theory of relativity or the theory of special relativity. This theory dispels the concept of absolute motion in **Newtonian mechanics** and instead proclaims only the relative motion between two frames of reference. Consequently, space and time are no longer viewed as separate and independent entities, but rather as a four-dimensional continuum called space-time.

spinor field

A **field** entity describing the **matter field** of spinning **quanta** is called a spinor field.

superconducting current

The minority of **quanta** constituting **matter** in the natural world manifest a specific characteristic of **quantum theory** discovered by Indian physicist Satyendra Nath Bose and then promoted by **Albert Einstein**. Namely, each **quantum state** (of motion) for those quanta can be occupied by any number of quanta, and extremely many quanta can be in the same quantum state. The quantum state (of motion) occupied by numerous quanta is called a **Bose-Einstein condensate**. If the quantum state maintains steady motion with nonvanishing, constant velocity and those quanta have electric charges, then this Bose-Einstein condensate will produce a specific macroscopic electric current free from resistance and damping. This electric current is called a superconducting current.

superconducting medium

Matter made of **quanta** in a **quantum state** of a **Bose-Einstein condensate** that manifests a **superconducting current** is called a superconducting medium.

superconductivity

A specific character of **matter** that realizes **superconducting current** is said to be superconductivity.

synchronicity

Carl Jung coined this term to indicate the acausal occurrence of a meaningful coincidence of events. He explained synchronicity in light of the **collective unconscious**.

T

theory of electromagnetism (Maxwellian electromagnetism)

James Maxwell proposed a theory of electromagnetism in terms of the **electromagnetic field** serving as a mathematical basis to explain the physical phenomena of electricity and magnetism in **classical physics**. The fundamental equation for the electromagnetic field, called the Maxwell equation today, is consistent with the **special theory of relativity**.

theory of special relativity
See **special theory of relativity**.

time derivative
The rate of time change is called the time derivative.

U

ultraviolet light
An invisible **electromagnetic wave** with **frequencies** ranging from about 10 to the power of 15 (i.e., 10^{15}) to 10 to the power of 18 (i.e., 10^{18}) is called ultraviolet light. Its **wavelength** is shorter than visible violet light and longer than X-ray.

V

vector
A physical or mathematical quantity described by an arrow placed in space is called a vector. Thus, a vector is a quantity specified by both the direction it points and its length. The length of an arrow is said to be the norm or the absolute value of a vector.

velocity
The time rate of the linear motion of **matter** in a given direction is called the velocity, and the speed if the direction is not specified.

velocity vector
A **vector**, that is, an arrow, representing both the direction and the magnitude of the time rate of linear motion is called a velocity vector.

visible light
The **electromagnetic wave** to which the human eye is sensitive is called visible light. Its **wavelength** ranges from 400 to 750 nanometers. One nanometer equals 10 to the power of -9 (i.e., 10^{-9}) meter.

von Neumann, Johann
A mathematician of genius born in Hungary in 1903, Johann von Neumann made remarkable contributions to the theory of games, the theory of **automata**, the theory of computing, the algebraic theory of operators, as well as the mathematical foundation of **quantum mechanics**.

W

wave

The transfer of **energy** by means of a regular vibration or oscillation is called a wave. A wave can occur either as the vibration of a material medium or as the variation of intensity of a **field**.

wavelength

The distance between successive crests or successive troughs measures the variation of intensity of a **field** or the vibration of a material medium when it reaches the maximum displacement; that measurement is called the wavelength of a wave. By definition, the wavelength is reciprocal to the **wave number**.

wave number

The number of crests or troughs lying on a line of one meter in length, or of any other unit length (such as one foot), in the transfer direction of a **wave** is called the wave number. It is also called the spatial frequency. The wave number, by definition, is reciprocal to the **wavelength**.

Y

Young, Thomas

English physicist, physician, and Egyptologist Thomas Young revived the **wave** theory of light and applied it to the **interference** of light.

Index

A

B

C

D

E

H

I

J

K

L

M

Q

R